D0408981
CENTRAL LIBRARY
828 "I" STREET
SACRAMENTO, CA 95814
JAN - - 2003

NEWTON

Also by Patricia Fara

Sympathetic Attractions:
Magnetic Practices, Beliefs and Symbolism in Eighteenth-century England

An Entertainment for Angels:
Electricity and Enlightenment

Patricia Fara

NEWTON

THE MAKING OF GENIUS

COLUMBIA UNIVERSITY PRESS
NEW YORK

COLUMBIA UNIVERSITY PRESS
Publishers Since 1893
New York Chichester, West Sussex

First published in Great Britain 2002 by Macmillan,
an imprint of Pan Macmillan Ltd, London

ISBN 0-231-12806-1

Copyright © Patricia Fara 2002

All rights reserved. No part of this publication may be reproduced, stored in or introduced into a retrieval system, or transmitted, in any form, or by any means (electronic, mechanical, photocopying, recording or otherwise) without the prior written permission of the publisher. Any person who does any unauthorized act in relation to this publication may be liable to criminal prosecution and civil claims for damages.

1 3 5 7 9 8 6 4 2

A complete CIP record is available from the Library of Congress.

Typeset by SetSystems Ltd, Saffron Walden, Essex
Printed and bound in Great Britain by
Mackays of Chatham plc, Chatham, Kent

For Michael

Contents

List of Illustrations

Every effort has been made to contact copyright holders of material reproduced in this book. If any have been inadvertently overlooked, the publishers will be pleased to make restitution at the earliest opportunity.

Acknowledgements

During the last few years, many people have contributed anecdotes and advice, but I should especially like to thank Anne Secord and Richard Yeo for their detailed critiques of draft chapters, and Jim Secord for his constant interest and advice, which included reading the complete final version. In addition, I am particularly indebted to Simon Schaffer and Judith Zinsser for their encouragement and comments, and I have also benefited from helpful discussions with Gadi Algazi, Malcolm Baker, Ulrike Boskamp, Michèle Cohen, Matthew Craske, Gideon Freudenthal, Cole Harrop, Michael Hau, Kilian Heck, Rob Iliffe, Ludmilla Jordanova, Milo Keynes, Nigel Leask, Christine MacLeod, David Money, Wendy Pullan, Steven Shapin, Skuli Sigardsson, Stephen Snobelen, Richard Staley, Ralph Stern, Jonathan Topham, Simon Werrett and Michael Wintroub.

For financial assistance, I wish to thank the Max Planck Institute, the Leverhulme Trust and the Royal Society; and for their assistance during publication, my thanks to my marvellous agent, David Godwin, and to my Macmillan editor, Anya Serota, who made many extremely helpful suggestions. I would never have completed this book without the invaluable support of relatives and friends, to whom I am deeply grateful.

Shortened versions of Chapters 2 and 8 have appeared as: 'Faces of genius: images of Newton in eighteenth-century England', in Geoffrey Cubitt and Allen Warren (eds), *Heroic Reputations and Exemplary Lives* (Manchester: Manchester University Press, 2000); and 'Isaac Newton lived here: sites of memory and scientific heritage', *British Journal for the History of Science* 33 (2000), 407–26.

Preface

> *Sublime spirit! Vast and profound genius! Divine being! Newton, deign to accept the homage of my feeble talents! . . . Surely even an idiot uses the same ink as a man of genius?*
>
> Étienne-Louis Boullée, 1784

Salvador Dalí's startling surrealist sculpture of Isaac Newton is an elegant abstract figure, its outstretched hand holding a ball on the end of a rope (*Figure 0.1*). Despite its rippling musculature, this polished bronze humanoid has a hollow body and a disturbingly empty oval instead of a face. By obliterating Newton's personality, Dalí implicitly invites us to impose our own interpretations. Similarly, generations of interpreters have created mythical visions of Newton from which the central core of the man himself is missing.

Although Newton wrote far more on alchemy, theology and ancient chronology than on either gravity or optics, he is now universally acclaimed as a scientific genius. Many good biographies fill in the details of Newton's life – Dalí's central void. In contrast, *Newton: The Making of Genius* examines how Newton was converted into the world's first scientific genius. The story of Newton's shifting reputations is inseparable from the rise of science itself. During the last three centuries, our views of Newton, science and genius have all changed dramatically, and this book explores these transformations. Repeatedly made to mean different things for different people, Newton has become an intellectual icon for our modern age, when genius commands the reverence formerly reserved for sanctity.

Newton was born well over 300 years ago, and much has happened since then. This may be stating the obvious, but it explains why comprehensiveness is not just impossible, but undesirable. To clarify the ways in which multiple versions of Newton's life have been created, this book deliberately leaves a lot out. It is emphatically *not* a conventional biography: on the contrary, one of its central arguments is that no 'true' representation of Newton exists. The narrative moves from Newton's lifetime to the present, hinging about the turn of the eighteenth and nineteenth centuries, a key period when science became consolidated and genius took on new meanings. Newton's ideas and opinions permeate this study of idolatry, but it is written for readers with no particular scientific, religious or historical expertise.

There are many different ways of telling history. History of science is a relatively new field, which came into its own after the Second World War. Partly in response to public repulsion at the atomic bomb, several eminent scientists wrote 'Plato to NATO' accounts that celebrated science's progressive march towards the truth. But these stories, appealing though they may be, now seem too simplistic and triumphal. Since the 1970s, sociologists have been minutely dissecting specific episodes from the past to reveal the social, political, economic and religious constraints that affect scientific practices and knowledge. Currently, historians are exploring new ways of incorporating these micro-studies within long-term analyses of science's rising power. This study of Newton's posthumous reputations responds to that challenge.

Newton is not just another dead white male scientist, but a major figurehead who symbolizes individual brilliance and scientific achievement. Moreover, he has helped to define what those very concepts mean. We can only view Newton's accomplishments and experiences through the refracting prism of a society that has itself been constantly changing. Examining his fleeting images illuminates how we have come to see ourselves.

1

SANCTITY

In Newton this island may boast of having produced the greatest and rarest genius that ever rose for the ornament and instruction of the species.

David Hume, *History of England*, (1754–62)

Borrowing the names of famous people does not necessarily bring good luck. During the nineteenth century, several young Isaac Newtons were prosecuted for forgery and other crimes, while French, German and American steam ships called *Newton* crashed on to rocks or burst into flames with alarming frequency. More recently, Apple has withdrawn its Newton range of computers, which failed to match up to expectations.[1] But other bearers of this illustrious name have been more fortunate: generations of Beatrix Potter fans have admired Jeremy Fisher's newt-like friend Sir Isaac Newton as he swaggered in his black and golden waistcoat, while the architectural writer Isaac Newton Phelps Stokes numbered among the wealthy American socialites glitteringly portrayed by John Singer Sargent. Images of the original Sir Isaac are ubiquitous, appearing not only on stamps throughout the world, but also in more specifically – if somewhat unexpected – British contexts, including Margaret Thatcher's coat-of-arms (*Figure 1.1*), the forecourt of the new British Library (*Figure 1.2*) and advertisements for the *Financial Times*.

Isaac Newton is now universally celebrated as a scientific genius, perhaps the greatest who ever lived. Yet Newton himself was not a scientist. Surprising as this assertion may seem, it is crucial for analysing his rise to glory. The word 'scientist' was not even invented until more than 100 years after his death, and Newton was an expert in fields that profoundly interested his contemporaries, yet have nothing to do with modern science. Unpaid, often mocked, his esoteric colleagues were as interested in moving nearer to God as in achieving progress towards a better world. Obsessed with alchemy, Newton constantly scoured the Bible for prophecies, redated ancient Egyptian chronology, converted his own mathematics back into the classical geometry of the Greeks, and spent thirty years chasing forgers as head of the Royal Mint in London.

'Does he eat, drink and sleep like other men?' inquired a French mathematician; 'I cannot believe otherwise than that he is a *genius*, or a *celestial intelligence* entirely disengaged from matter.'[2] Often retold, such anecdotes contributed to Newton's canonization as a secular saint endowed with supra-human capacities. Not everyone regarded Newton with such esteem, however. When the unknown and reclusive Cambridge scholar first appeared on the philosophical stage he was strongly criticized, and sceptics continued to launch virulent attacks right through the eighteenth century. Newton has frequently been accused of mental instability or even insanity, his scientific theories have been constantly reinterpreted or even rejected, and the overriding goal of his studies was to learn more about God.

How, then, did Newton become world famous as a brilliant scientist? The obvious answer, that he discovered fundamental laws of nature, is too simple. For one thing, philosophers question whether scientific knowledge can ever be absolutely and permanently true. But even without venturing into these huge debates, it is clear that Newton's legacy is problematic. We often talk about a Newtonian world view, but that term is deeply ambiguous, since Newton's successors interpreted his ideas in directions that he would find unrecognizable. Moreover, in the early twentieth century, Albert Einstein showed that Newton's ideas were of little help in describing the quantum world of sub-atomic particles.

Newton's centrality in theoretical physics may have been

displaced, but his legendary reputation endures. Recent biographers have portrayed Newton as an alchemical and biblical expert convinced of God's presence throughout the universe, yet he still symbolizes the committed scientist emotionlessly investigating a mechanistic world. Rather than searching for more facts about Newton himself, this book explores how he became celebrated as a national hero and a scientific genius – a secular saint for our modern society.

Matters of fact

Even the briefest survey of Newton's life unsettles his image as the idealized prototype of a modern scientist.[3] Like many of his contemporaries, Newton was engaged in a wide range of activities, many of which fell far beyond the scope of what we would expect of a scientific figurehead. A renowned expert on Jason's fleece, Pythagorean harmonics and Solomon's temple, his advice was also sought on the manufacture of coins and remedies for headaches. On the other hand, he was free of the responsibilities besetting today's international high-fliers. Newton had no laboratory team to supervise, no obligation to generate commercially viable research projects, and never travelled outside eastern England – his most adventurous journey was a trip up the Thames to the Astronomical Observatory at Greenwich.

This reluctance to travel provides a useful framework for recounting Newton's life in three phases, corresponding to the three places where he lived – Lincolnshire, Cambridge and London. Such a geographical approach, apparently based on well-established facts, conveys a reassuring ring of historical truth. But, of course, even the most apparently straightforward biography is structured according to its author's beliefs. Exploring Newton's posthumous existences entails confronting a fundamental historical problem: circularity. To appreciate the diverse images of Newton that were created after he died, it is essential to have some basic knowledge of what are generally accepted as facts about his life and work. Any attempt to present such information neutrally is impossible, however, since each biographer will have a different view of what is important. Even worse, any discussion of how Newton has been portrayed in the past must itself enter into the archive of Newtonian representations, and so affects how he will be viewed in the future. By analysing the processes through which myths are made, this book is itself altering their interpretation. All that can be presented is one version of the 'facts' of Newton's life and achievements . . .

Born in a small Lincolnshire hamlet in 1642, Newton was brought up mainly by his grandmother until he was twelve, when he was sent away to the nearby market town of Grantham to attend the local grammar school. With only a brief interlude back home, at the age of eighteen Newton went to Trinity College, Cambridge, where he remained for most of the next thirty-five years. As a student, he subsidized his meagre allowance by performing menial chores and initiating a money-lending enterprise. Although the examination system was mostly a formality, he did dutifully broach the officially prescribed Aristotelian texts. But Newton also explored extra-curricular books on history, astrology and modern European philosophy, teaching himself the mathematics he needed to understand the novel ideas being put forward by controversial scholars such as René Descartes, the French natural philosopher.

By the summer of 1665, after four years of intensive and self-directed study, this solitary scholar had made little impression on his colleagues. There are no recollections of him by other students, and Isaac Barrow, the Lucasian Professor of Mathematics (a position that he later handed over to Newton), 'conceived then but an indifferent opinion of him'.[4] But Newton's life suddenly changed when he retreated to Lincolnshire for about eighteen months to escape the plague sweeping through Cambridge.

The year 1666 became celebrated as Britain's *annus mirabilis*, when the nation's fleet triumphed over the Dutch, and London survived the Great Fire. Newtonian historians have described

1665–6 as Newton's personal *annus mirabilis* when, forced into rural retirement, he compiled a staggering array of new mathematical and scientific techniques. Half a century later, Newton boasted (perhaps a touch wistfully) that 'in those days I was in the prime of my age for invention & minded Mathematicks & Philosophy more than at any time since'.[5]

This was when Newton supposedly gained inspiration by watching an apple fall from a tree, and biographers often depict an Arcadian interlude of frenetic and almost overnight creativity. Nevertheless, such tempting tales ignore the long periods Newton dedicated to experimental and theoretical confirmation of his theories. Moreover, some of the dates inconveniently refuse to comply with this simplified picture. Effectively exiled into academic solitary confinement, Newton did not immediately and single-handedly revolutionize the seventeenth-century scientific world with the fruits of his research. However, it is fair to say that he made key discoveries in mathematics, optics and dynamics, which formed the foundation for much of his own subsequent work, and affected the future course of science.

Once back in Cambridge, Newton adopted a solitary life, and spent much of the next two years secretly poring over alchemical manuscripts and experiments. He was shocked out of this seclusion in 1668, when a new book on mathematics forced him into print to establish his own priority, and he was soon appointed the Lucasian Professor of Mathematics. Although Newton was to hold this post for thirty-two years, he was a poor lecturer who often 'for want of Hearers, read to ye Walls',[6] and he increasingly neglected his teaching duties. Immersed in his research, he was only interested in communicating his ideas to other mathematical experts.

Yet Newton's first public success was not with a new theory, as his subsequent reputation might lead us to expect, but with a small reflecting telescope that he built himself, even grinding the lenses by hand. Only 15 cm long, Newton's telescope could magnify distant objects far more powerfully than larger models, and in 1672 he was elected to the Royal Society. In his first lecture, he presented many of the ideas that would overturn not only the science of optics, but also the methodology of scientific practices. Subsequently developed into the *Opticks*, one of his most famous books (first published in 1704), Newton's early accounts of his experiments with prisms simultaneously rewrote the nature of light and set theoretical work on a new experimental basis.

Newton described to the Fellows what is often called his crucial experiment, in which he used two prisms to demonstrate that sunlight is composed of coloured rays of light (*Figure 1.3*). He aimed to reject the prevailing view, which was essentially Descartes's reworking of Aristotelian ideas, that the colours we see around us occur because white light is modified when it interacts with an object's surface. Newton argued that different colours are inherently present in sunlight. Conceiving light as streams of tiny particles that are slowed down when they pass through glass, he explained that a prism separates light out into its constituent coloured rays.

In this early work on optics, Newton also laid the basis for his experimental approach, which profoundly affected the ideology of scientific research. The way forward, he insisted, was not to devise abstract hypotheses, but to build theories on the twin pillars of mathematics and experiment. That this does not now seem such a revolutionary suggestion is precisely because Newton's innovations have become fundamental principles of modern science. But before then, geometry, experimentation and natural philosophy had been three distinct domains on the map of knowledge, traditionally occupied by people with different skills and goals. Henceforth, preached Newton, theories would be the consequence of observations, not their inspiration.

Far from following up on his controversial entrée into the international world of natural philosophy, Newton withdrew into Trinity College and devoted much of the 1670s to pursuing alchemy and theology. He was also absorbed in mathematics, an aspect of his work that tends not to receive much attention, perhaps because people find it difficult. Some of Newton's conclusions proved extremely influential, particularly the neat formulae he derived for curves and series of algebraic expressions. Another significant innovation, which he called fluxions and we call calculus, has become particularly famous because it led to a bitter priority row between Newton and his arch-enemy, the German mathematical philosopher Gottfried Leibniz. Their successors energetically perpetuated this international dispute for decades, and historians are still finding fresh perspectives from which to analyse it.

At the same time as developing new mathematical techniques, Newton was scouring books and manuscripts to compile information about ancient chronology, religious doctrines and biblical prophecies. Owing to his belief that orthodox interpretations of Christ's holy status were wrong, Newton received a royal exemption from the normal obligation for Cambridge Fellows of being ordained in the Anglican Church. Convinced that scholarly interpretation could restore the original meaning of corrupted scriptural texts, he also sought to retrieve arcane alchemical knowledge. This was no mundane search for the philosopher's stone or the elixir of life, but a quest of the soul. Newton believed that a divine vegetative spirit pervades the world and effects material and spiritual transformations, governing changes in metals as well as the growth of plants and animals. Converting a College garden shed into his private alchemical laboratory, he constructed his own furnaces to explore in secret these processes of natural development. Newton continued this research until the mid-1690s, and his published works on gravity and optics – those now seen as the foundation of modern science – are suffused with alchemical and religious concepts.

In the early 1680s, a series of comets blazed across the sky, arousing terrified fascination throughout Europe. Many people interpreted these celestial spectacles as prophetic messages from God, and Newton became obsessively interested in these unpredictable phenomena. Spurred on by discussions and correspondence with his associates, he dedicated himself to mathematical astronomy and started writing his most famous book, the

Philosophiæ Naturalis Principia Mathematica (*Mathematical Principles of Natural Philosophy*). First published in 1687, and twice revised to accommodate criticisms, the *Principia* lies at the heart of Newton's subsequent reputation because it provided a new cosmology.

Even though we may not realize it, we view the universe through Newtonian spectacles. This makes it hard for us to imagine older ideas and take them seriously. Newton was born at a time when traditional views still survived. Some people were still arguing about the displacement of the earth from the centre of the planetary system, and Newton himself was affected by the Aristotelian distinction between the earth, which is constantly in flux, and the unchanging heavens, which rotate in divinely perfect circles.

Conflicting theories had been put forward during the seventeenth century. One influential model was proposed in 1600 by the English physician William Gilbert. In his cosmos, the sun and the planets are bound to each other magnetically (which is why the poet John Milton referred to the sun's 'magnetic beam' in *Paradise Lost*). It was Gilbert's magnetic beliefs that directed the research of Johannes Kepler, whose demonstration that planets move in elliptical orbits crucially affected Newton's own work. Other natural philosophers, most notably Descartes, objected to the idea of an invisible occult force extending its powers as if by magic. Descartes insisted that action depends on contact, so his universe is packed with tiny particles that push against each other and swirl around in patterns called vortices.

The *Principia* revolutionized the course of physics by providing a single mathematical law to describe the motion of heavenly bodies as well as minute particles of matter on earth. For the first time, natural philosophers could provide reliable forecasts of when a comet would reappear. This helped them to claim that their approach to the world was superior to astrological or biblical predictions, and thus to wrest authority from traditional experts. That a complete manuscript ever reached the press was largely due to the persistent persuasion of Edmond Halley. Although then merely the paid Clerk of the Royal Society, Halley later became famous in his own right as the Astronomer Royal who correctly forecast the return of the 1682 comet that now bears his name. For Newton also, this research into comets lay at the heart of his subsequent fame.

Written in Latin and packed with geometrical diagrams, the *Principia* appears a dry book, but for those who understood it, Newton wrote in a persuasive style. Right at the beginning, he stated his three laws of motion, which govern how objects move and interact with one another. Most people first encounter these laws at school, when asked to solve problems about colliding billiard balls, or lorries rolling down hills. Newton's great coup was to apply these laws to describe the motion of the planets, thus uniting events on earth with motion in the heavens. He introduced a new concept of gravity, picturing a universal attractive force stretching out through space, one which affected comets, falling apples and tiny atoms in the same way. Unlike Descartes, Newton visualized large tracts of empty space not only between the heavenly bodies, but also between the particles that make up apparently solid matter.

Just as importantly, Newton expressed gravity's effects mathematically. The nearer to one another two objects are, and the heavier, the more strongly they attract each other. This is known as the inverse square law, because this attractive force depends on the square of the distance between the objects. While Albert Einstein is celebrated for the equation $e=mc^2$, so Newton's work is symbolized by the $1/r^2$ relationship.

Newton's book also abruptly altered the pattern of his own existence. In addition to the deluge of congratulations, criticisms and controversies, other events were forcing Newton to reappraise his life. In particular, with the departure of his friend Fatio de Duillier, a young Swiss mathematician, the only close adult relationship he ever formed came to an end. A few weeks later, in the autumn of 1693, after he started sending bizarre letters to his colleagues, rumours circulated that he had gone mad or even had died. Becoming even more reclusive, Newton turned in on himself, continuing his alchemical experiments and revising his manuscripts.

In 1696, Newton emerged from this self-imposed seclusion and embarked on a totally new career at the Mint. Enjoying metropolitan prominence, he became England's most celebrated and powerful natural philosopher. As Warden and later Master of the Mint, he pursued his duties with an intensity matching his previous devotion to alchemy, theology and mathematical astronomy. He instituted major reforms, and zealously persecuted fraudulent money-makers – even to the extent of arranging their executions.

Elected President of the Royal Society in 1703, Newton became an authoritarian patron and administrator, ensuring that his influence and his ideas extended throughout Europe. The following year he published the first edition of the *Opticks*. Although its ideas were no longer controversial, this book comprised a manifesto presenting his mathematical, experimental style of research. As successive editions appeared, Newton added an increasing number of speculations about fundamental topics such as the nature of matter and its relation to life. Disguised as 'Quæries', these ingenuously phrased speculations often contradicted his earlier ideas, and formed the experimental agenda for his eighteenth-century successors. Responding to critics, Newton also revised the *Principia*, in 1713 adding an appendix (called the General Scholium) that emphasized God's constant presence throughout the universe. As before, theology and natural philosophy were inextricably linked together.

From his knighthood in 1705 through to his death in 1727, Newton continued working at the Mint. At the same time, he was actively involved in the international community of natural philosophers, rewriting and publishing earlier work in mathematics, optics and astronomy, and supervising his vicious priority dispute with Leibniz. But in private, his major concern was to consolidate his previous theological studies. Juggling with dates to reconcile conflicting events and opinions, Newton endlessly revised his manuscripts on ancient chronology and biblical prophecy. Shortly after he died, sanitized versions that effectively concealed his heretical religious ideas were published. His heirs had already put into motion the machinery designed to protect and enhance his reputation.

A secular saint

Like William Shakespeare, England's other most exalted genius, Newton's reputation has been repeatedly refashioned.[7] Indeed, it is precisely because his life has been constantly reinterpreted that we can examine how he became converted into a national scientific hero. Even such a basic fact as his date of birth is unclear. England was then ten days out of step with most of the rest of Europe. Ironically, Newton would himself urge the government to reform the English calendar, but it was not until 1752 that the country belatedly moved out of its self-imposed isolationism. So although Newton was born on Christmas Day 1642 in England, in France and other European countries it was already 4 January 1643.[8]

Different types of uncertainty shroud other aspects of Newton's life. Although it is common knowledge that he watched an apple fall from a tree, historians continue to argue about the significance of this celebrated event and indeed whether it occurred at all. We remain uncertain about his appearance, since contemporary descriptions and portraits give conflicting pictures. Was he a thin, prematurely grey scholarly type with a piercing gaze (as in *Figure 2.1*), or was he a plump, brown-haired man with a distant demeanour (*Figure 2.2*)? Looking back, other large question-marks hang over his life. Did he experience a period of insanity from which he never fully recovered – and if so, was this an inherited problem, or one brought on by overwork or experimenting with dangerous chemicals? Did he turn a blind eye to his niece's clandestine love affair in order to gain his powerful post at the Mint? And what about his own love life – did he renounce romance for science, did he enjoy homosexual relations with younger men, or was he emotionally damaged by his father's death before he was born and by his mother's remarriage when he was three years old?

Over the last 300 years, Newton's biographers have argued about the answers to these and many other questions. They have disagreed about his major achievements, and what significance he attached to different aspects of his work. Were his alchemical ideas central to his cosmological theories, or were they the embarrassing delusions of an otherwise supremely rational intellectual? Should we regard his long years at the Mint as the patriotic duty of a dedicated administrator, or the government's exploitation of an underpaid academic? Do his theological books comprise the sad ramblings of an elderly man, or do they confirm a lifelong religious commitment?

Although researchers are still uncovering new details, examining such issues is made more difficult by the absence of manuscripts that have been destroyed over the years by enthusiasts eager to preserve Newton's public reputation. Still more importantly, all Newton's biographers have selected from the vast corpus of available information only what they feel to be relevant facts. They have disagreed not only over what these facts are, but which ones are significant. There are several reasons for these differences in approach. Partly they reflect trends in historical fashion. Compared with the Victorians, for instance, modern writers are more inclined to integrate a famous subject's personal and public lives, and to show how emotional and social experiences are inseparable from achievements, whether these be scientific discoveries, military victories or philosophical inquiries. Furthermore, writers obviously tailor their descriptions of Newton to suit their readers. Thus one might expect (not always accurately, as it turns out) an entry in a children's encyclopaedia to include more information about Newton's own childhood than the introduction to a scientific textbook.

But changes in Newtonian biography also reveal more specific transformations. Understanding how Newton has become a cultural icon entails not just studying Newton himself, but also examining how society's attitudes towards science, famous people and fame itself have changed during the last 300 years. Authors have created various versions of Newton's life because they have held different views of what it means to be a successful person. There is no simple one-way relationship between what society at large judges to be the characteristics of greatness, and the biographical accounts that are produced. These biographies themselves help to formulate who is famous and how famous people are defined. Thus the shifts in Newton's reputation have simultaneously mirrored and moulded broader social perceptions.

None of Newton's contemporaries shared our view of him as a 'scientific genius', because that concept had not yet been invented. Countless representations of Newton have themselves contributed to our understanding of what the terms science and genius mean. The past is often said to be a foreign country, and words such as science and genius are deceptive, because their meanings have repeatedly altered.[9] In Newton's time science meant something resembling systematic knowledge, so that although Newton's experimental colleagues were called natural philosophers, only some aspects of their activities came to form the antecedents of modern science. Natural philosophy was an umbrella term embracing different practices, but its major objective was to learn more about God through studying the natural world. In a widely used phrase coined by the chemist Robert Boyle, one of Newton's associates at the Royal Society, these new 'priests of nature' read and interpreted the divine book of the natural world rather than God's other great book, the Bible.

The wealthy gentlemen who studied and experimented in the privacy of their own homes or university studies fervently believed in the value of their research, but they enjoyed little public interest or government support: indeed, they were often viciously caricaturized. Among the frequent satires that mocked the pretensions of gentlemanly collectors and opportunistic inventors, the most famous example is now Jonathan Swift's *Gulliver's Travels*, first published in 1726. Swift parodied the Royal Society as the Academy of Lagado, staffed by bumbling professors turning the cranks of unworkable language machines, and frequented by unrealistic schemers trying to make cucumbers out of sunbeams.

Newton is often listed first among influential scientists, but in his own lifetime relatively few people had heard of him and science as we know it simply did not exist. It was not until the early nineteenth century that scientific inquiry started to achieve its current prestige. Science and Newton's fame grew together and fed on each other. As science became valued for its contributions towards commercial, industrial and military prowess, new scientific disciplines with their own specialized societies proliferated. Choosing to become a man of science slowly started to become a genuine career option.

The word 'scientist' was coined only in 1833, and was not widely used until the end of the nineteenth century. Even Charles Darwin, who lived until 1882, never referred to himself as a scientist. Reinforcing Newton's intellectual reputation provided one way for scientific propagandists to advertise the value of their work. Promoting Newton as an English paragon of achievement was tied up with the swelling importance of science, as well as with chauvinistic claims that England was the world's leading scientific nation. Long after Scotland and England had been united, Newton was hailed as an English rather than a British hero: even the Scottish inventor James Watt was sometimes described as an Englishman.

Genius is the second slippery word that simultaneously characterizes and confuses Newton's rising renown. At the end of the eighteenth century, the German philosopher Immanuel Kant definitively declared that Newton was not a genius. He made this startling statement not because he was unimpressed by Newton's achievements (on the contrary, he idolized the creator of such 'immortal work') but because he was, like many of his contemporaries throughout Europe, interested in analysing the character of genius itself.[10] Nowadays writers use the term genius very liberally, and a book using the word in its title appears every couple of weeks. Kant's counter-intuitive assertion underlines how important it is to investigate more closely the concept of genius and how it has altered. Far from being an objective term, genius is a tribute with no permanent definition. One reason why we unhesitatingly include Newton among the world's greatest geniuses is that people's changing perceptions of Newton over the past 300 years have themselves affected what it means to be a genius.

For much of the seventeenth and eighteenth centuries in England, genius was regarded as a quality possessed in varying amounts by all talented people, a gift from God that enabled them to excel in a particular field. Newton was not generally celebrated as an individual genius, but was more often singled out as being blessed with a particular genius for mathematics. Towards the end of the eighteenth century, however, the German Romantic writers clustered round Johann Wolfgang von Goethe gave the word a new meaning. Genius gradually became the label not for a human characteristic, but for a singular man (significantly *not* a woman) who – like the mad – was set apart from the rest of society.

Romantic geniuses displayed the asceticism formerly attributed to saints, but their flashes of inspiration were said to come not from God, but from an internal creative urge. Newton, the personification of abstract reason, became the first exemplar of a scientific genius. This new label was paradoxical to the point of being oxymoronic. Creative Romantic geniuses regarded themselves as being governed by their passions, and they searched for knowledge within themselves. Men of science, on the other hand, were prized for their detached rationality and their objective analysis of the external world. As Romantic authors attempted to reach beyond individual ability to articulate an ungraspable absolute, they imbued genius with an aura previously reserved for religious ideologies. In the twentieth century, while physiologists tried to give genius a material existence by locating it in brain and nerve anatomy, psychometricians were introducing new IQ vocabulary to quantify human characteristics. These scientific searches for laws governing the incidence of genius assumed that such a transcendent notion could be objectively analysed.[11]

One way of thinking about these transformations in both science and genius is to regard them as aspects of the long-term secularization of society. While science was becoming publicly recognized as a valuable activity, people were turning away from the Bible and looking towards scientific endeavours for information about their world. The miraculous powers previously attributed to holy figures were transferred to scientists, the new saviours of sick bodies and a deteriorating environment. Genius resembles sanctity, both words being impossible attempts to pin down an ineffable quality. Even the word itself has divine associations, since one of its classical roots refers to the small spirit or *genius* supposed to accompany a man through his life.

Natural philosophy and theology shared a common cultural and intellectual heritage, and polemical scientific writers have frequently drawn on religious imagery of disciples and sects to advertise their allegiances. The students of the Swedish botanist Carl Linnaeus named themselves his apostles, and exhibited missionary zeal as they successfully converted the world to his system for classifying plants and animals. Echoing Boyle, eighteenth-century 'priests' lectured on the 'religion of nature', while Victorian scientists frequently celebrated Newton as the 'high priest of science'. John Conduitt, one of Newton's most ardent hagiographers (to whom we owe the survival of many personal details), enthused that his 'virtues proved him a Saint & his discoveries might well pass for miracles'.[12]

While smacking of hyperbole, such descriptions are not just effusive accolades, but reflect the vital role of saints as moral exemplars who embody cultural ideals. Scientific organizations have gradually taken over social roles that were formerly played by religious institutions. To describe Newton as a secular saint implies not that he lived a pure and holy life, but that people have invested – and continually re-invest – his image with values that they hold to be centrally important.

Like role models in secular spheres, over the centuries saints have fulfilled different functions. The Church's first martyr, St Stephen, exemplified how an ordinary mortal could imitate

Christ's behaviour at the Crucifixion. Horrific pictures showing him meekly being stoned to death were originally intended to inspire Christian forgiveness, even though they are now hung not in religious centres but in art galleries, the secular temples of our time. The criteria for beatification gradually changed, and subsequent saints were often canonized for their symbolic value rather than their lives. Thus the Vatican now seems concerned to demonstrate its international democratic ideals by selecting for canonization lay people who a couple of centuries ago would not have been considered appropriate candidates.[13]

Science and religion, genius and sanctity: from some perspectives they seem completely distinct from one another, yet their cultural roles are closely related. As Western society has become increasingly secularized, intellectual ability has gradually displaced saintly dedication as an attribute of greatness, while medical and technological experts have assumed the mantle of miracle workers. Enlightenment England converted Westminster Abbey into a secular shrine commemorating the achievements of the nation's great men, and it is here that Newton was buried. This was one of the first steps in making Newton into a secular saint for the twenty-first century.

Interpretations

Newton's *Principia* was not expected to be a best-seller. The Royal Society declined to back it, since their finances had just been exhausted by an expensive but unsuccessful *History of Fishes*, so Halley bore the publication costs himself. In 1687, only three or four hundred copies were produced by Joseph Streater, a printer who anticipated far more profit from the pornographic literature for which he was subsequently imprisoned. Twenty years later, original copies were so rare that they could fetch almost five times their original price.

Newton deliberately made his book accessible only to a privileged, knowledgeable elite. As he later explained to a colleague, 'to avoid being baited by little Smatterers in Mathematicks . . . he designedly made his Principia abtruse; but yet so as to be understood by able Mathematicians'.[14] Newton may have attempted to restrict his audience, but he and his allies had already ensured that his renown as a brilliant if eccentric mathematician had seeped out beyond the walls of Cambridge. A select international circle of learned natural philosophers eagerly snapped up his long-awaited text, although – like the philosopher John Locke – many of them admitted skipping the more erudite chunks of mathematics.

However, even among scholars, the *Principia* was far from being an immediate and unqualified success. Some of Newton's contemporaries condemned his concept of gravitational attraction for being a fancy term that explained nothing, 'a Simile or rather a Cover for Ignorance', as the eminent lawyer Roger North sneered. Like many of Newton's educated contemporaries, North was never converted away from Descartes's model of the universe, in which particles interacted by direct contact rather than through attraction at a distance. North voiced the views of many when he commented sarcastically, 'If one asks why one thing draws another – It is answered by a certain drawingness it hath.'[15]

Because the *Principia* has now become a seminal work in the history of physics, it is hard to recognize that, like many scientific works, its huge influence was not simply due to the ideas it contained. Looking back, it is relatively easy to find ample references hymning Newton's glory. What is more challenging is to retrieve other ways in which Newton and his works were perceived by his contemporaries, and to explore how his personal reputation, as well as the meaning of Newtonianism, have constantly changed.

One immediate and surprising conclusion is to realize that Newton was nowhere near as famous in the first quarter of the eighteenth century as he is now. While he was certainly well known among scholarly circles of natural philosophers, most English people (let alone foreigners) had never heard of him. Esoteric academic theories held little relevance for the problems of daily life, and there were as yet few popular journals or encyclopaedias containing scientific information at affordable prices. Newton was absent from places where we might expect to find him, such as collective biographies of England's great men or university reading lists. His relative insignificance was partly due to the low status of natural philosophy, which had not yet become a standard component of educational curricula. Most learned gentlemen were far more concerned with theological and literary texts, so that catalogues of important English writers regularly included names now only dimly remembered.

England's other major intellectual hero during the eighteenth century was Alexander Pope. Although he was often satirized, this celebrated poet and essayist conformed with Enlightenment ideals of gentlemanly achievement, which leaned heavily on the models provided by Greek and Roman culture. In addition to being prized for his own witty learning, Pope was associated through his translations and editorial work with the two traditional examples of literary genius, Homer and Shakespeare. As scientific activities gained greater prestige, Newton gradually became paired with Pope, so that they became twin icons of English greatness modelled on a classical inheritance. The eminent collector Richard Mead, who was also Newton's personal physician, hung portraits of Newton and Pope next to one another, 'near the Busts of their great Masters, the antient Greeks and Romans'.[16]

Members of the burgeoning middle classes believed that their surroundings should be educationally improving as well as aesthetically uplifting, and foreign visitors frequently commented on the English custom of decorating homes with commemorative busts and pictures of their national heroes. When the Dutch Philipse family wanted other American settlers to recognize their English affiliations, they redecorated their Manhattan home to emulate the latest London fashions, including an elaborate rococo ceiling embellished with busts of Pope and Newton. Such grandiose schemes of refurbishment were undertaken to display an owner's wealth, patriotism and sophistication, but they simultaneously reinforced Newton's growing reputation. Repainting the ceiling may seem a strange way of celebrating the innovator of a new theory of the universe, but Enlightenment cultural figureheads were fashioned in drawing rooms and coffee houses. In the absence of any formal structures for scientific education, people learned about Newton from chatty journal articles or reading poetry, then an enormously popular genre for transmitting ideas.

Debates about religion and philosophy were saturated with national interests, and patriots who knew little about optics or gravity fêted Newton because he was an Englishman. Despite claims that the bonds of international scholarly brotherhood transcended geographical borders, natural philosophers were greatly affected by current political affairs. France and England were almost constantly at war during the eighteenth century, which gave an added edge to the rivalry between Descartes and Newton. John Desaguliers, one of London's leading lecturers, chauvinistically emphasized that Newton had established English supremacy on the scientific battlefield: 'It is to Sir *Isaac Newton*'s Application of Geometry to Philosophy, that we owe the routing of this Army of *Goths* and *Vandals* in the philosophical World.' Similarly, Newton's protracted argument with Leibniz attracted antagonists on both sides because it had more to do with the Hanoverian succession to the English throne than the relatively esoteric question of who had invented calculus first. Newton himself listed achievements by their country of origin, and discreetly asked travellers to act as industrial spies by bringing back useful commercial information about foreign inventions.[17]

By the end of the eighteenth century, Newton had been incorporated within a pantheon of great Englishmen (along with the occasional woman, particularly Elizabeth I). These heroic figures were bonded by their nationality and their timeless achievements, and included famous characters from the past as well as Newton's contemporaries, notably Locke. At this time, scientific and medical practitioners were strengthening their identity by demonstrating that they could contribute to the country's financial and social welfare. For them, it was vital that Newton be simultaneously a scientific and an English hero.

The idea of genius was also intimately tied up with chauvinistic ambitions. British aesthetic philosophers often compared French rule-bound poetry unfavourably with English creativity. In 1759, the poet Edward Young published a long essay on originality that strongly influenced writers on genius during the Romantic period. For once boasting about Britain rather than England, Young declared that '*Bacon*, *Newton*, *Shakespeare*, *Milton*, have showed us, that all the winds cannot blow the *British* flag farther, than an original spirit can convey the *British* fame.'[18]

Newton's escalating fame and the expanding importance of natural philosophy were closely linked with England's explosive commercialization during the eighteenth century. The economy boomed as customers displayed new patterns of spending, eagerly purchasing replaceable consumer products like Josiah Wedgwood's pottery plates or Matthew Boulton's metal ornaments. Entrepreneurs primed with the latest research into scientific and retailing techniques energetically marketed instruments, texts and lecture courses, and knowledge of the most recent scientific discoveries became an essential component of genteel education.

This twinned fascination with deciphering the world's mysteries and exploiting its resources was captured by the Midlands artist Joseph Wright, whose incandescent pictures of erupting volcanoes resemble his sparking iron forges and glowing factories. Wright's picture of *An Experiment on a Bird in the Air Pump*, widely reproduced on modern book covers, has come to symbolize science's ambiguous attraction. As children turn away in horror from a bird trapped inside an evacuated glass dome, the natural philosopher with his hand on the air tap seems about to choose whether the bird should live or die. Wright's picture comments on the new control that experimenters exerted over both nature and society. Whereas astrologers had formerly been the experts in cometary behaviour, Newtonian natural philosophers gained prestige by boasting about their accurate astronomical predictions: scientific innovations are about power as well as knowledge.

In Wright's companion picture, a domestic group clusters enthusiastically round an orrery, a large mechanical model of the planets' movements about the sun (*Figure 1.4*). Named after an early purchaser, the Earl of Orrery, orreries became a standard piece of equipment for travelling lecturers who brought Newtonian astronomy to households and lecture theatres throughout Europe. The more expensive models were several feet across, decorated with brass horses' heads and incorporating finely tuned mechanisms to ensure that all the known planets – with their attendant moons – rotated at appropriate speeds when the experimental performer turned a handle. (Orreries can be dated by the number of planets, since Uranus – initially named Georgium Sidus after George III – was not discovered until 1781.)

In Wright's orrery, a candle represents the sun, giving a central illumination deliberately reminiscent of Dutch religious imagery. The spectators' faces reflect this light at different angles, progressing round the picture like the phases of the moons and planets. Orreries emblematized the orderly Newtonian cosmos as well as stable, hierarchical British society: just as physical bodies gravitate towards one another, so too the members of Georgian society were bonded together by ties of sympathy, ranging from the intimate affection between the two children to the experimenter's patriarchal benevolence. By the time John Keats went to school, teachers were inculcating Newtonian regularity in recalcitrant pupils by training them to simulate planetary motion in giant 'living orreries' stretching across the playground. Imagining his weekly thirty minutes slowly circling with a card showing his planetary identity gives new significance to Keats's famous lines from 'On First Looking into Chapman's Homer':

> *Then felt I like some watcher of the skies*
> *When a new planet swims into his ken . . .*[19]

Enthusiastic scientific propagandists seeking to attract larger audiences fashioned Newton's public image and gave natural philosophy a new direction. Their books and lectures incorporated Newton's insistence that natural laws were to be found not from studying books or framing speculative hypotheses, but by bringing the power of pure mathematics to bear on observations and the results of practical experiments. Nevertheless, adopting this mathematical experimental approach to the world supplemented rather than displaced religious quests for knowledge. As a journal article explained, Newton 'shew'd that the World was *philosophically* and *mathematically* made: and that it could be *framed* and *held together* by none but an *infinitely wise* and *Almighty* Architect'.[20]

Writers also used Newton's name to establish some of the social norms that have influenced modern science. For one thing, scientific observation and thought were only suitable activities for men. Journalists instructed their readers that 'the most beautiful Woman in the World would not be half so beautiful, if she was as great a Mathematician as Sir *Isaac Newton* . . . Learning is so far from improving a Lady's Understanding, that it is likely to banish the most useful Sense out of it.' Although popularizers were marketing versions of Newton's philosophy designed for women, their simplified little books reinforced views that women should be attentive recipients of scientific knowledge rather than active participants in research.[21]

But even though women were being coopted as audiences rather than practitioners of natural philosophy, the rational sciences were often portrayed as female muses. The frontispiece of the first English edition of the *Principia* shows the goddess of mathematics communicating her wisdom to a semi-divine Newton (*Figure 1.5*). Nature frequently appeared as a scantily clad woman whose hidden secrets could be unveiled by male experimenters. One anonymous author verbally reinforced this ideology: '*Newton* was eminently distinguish'd by his *deep Searches* into Nature herself; He was Nature's Son: He seem'd to understand all her Mysteries, and to be sent into the World on Purpose to *lead Mankind* into the highest Notions of the Wisdom, Goodness, and Power of the *Great Author* of Nature.' By portraying Newton as a secular saviour, the son of a female nature, such assertions helped to transfer the location of spiritual knowledge about God from the Bible to the physical world. Newton stood, the writer continued, 'at *the Head* of Philosophy and Mathematicks, wherever *Learning and Knowledge* have expanded their Empire'. Just as political rulers and commercial entrepreneurs were establishing the nation's imperial might, Newton was being elevated into an intellectual leader who would ensure England's cultural dominion.[22]

Eulogies frequently blurred the distinctions between Newton as originator and transcriber of God's designs, so that Newton himself acquired a semi-divine status. Edmond Halley set the tone with the Latin poem he composed to preface the *Principia*, which hymned Newton for being able 'to penetrate the dwellings of the Gods and to scale the heights of Heaven'. Halley's lines inspired the frontispiece of the first English translation of the *Principia*, published two years after Newton had died (*Figure 1.5*).[23] Playing on a favourite Enlightenment visual pun, rays from a divine intellectual sun disperse the dark clouds of ignorance. In this secularized reworking of traditional religious iconography, Newton's heavenly pose recalls that of a saint receiving divine inspiration not from an angel, but from the Goddess of Mathematics, identified by her dividers. Even though Newton's planets move in ellipses, their orbits here are circular, a traditional rendering of cosmic perfection.

A decade later, Newton was even more explicitly deified in the frontispiece of Voltaire's hugely successful if somewhat idiosyncratic version of Newtonian optics and gravity (*Figure 1.6*). Newton himself now measures the globe, and his head lies close to the source of the heavenly light reflected by Truth's mirror on to Voltaire, the earthly scribe below. Similar imagery appeared in one of English literature's most famous frontispieces, William Blake's *The Ancient of Days*. Using striking yellows and reds, Blake showed his Newtonian God Urizen emerging from a cloud that flames out of the void, stretching down one arm to demarcate the world with his mathematical dividers, the biblical instruments of creation.[24]

As Newton became more famous, natural philosophers recognized the advantages of allying themselves with his name. Labelling oneself a Newtonian came to command instant respect, and by the middle of the century, English natural philosophy was ruled by a Newtonian orthodoxy that was hard to contradict. Books challenging Newtonian concepts were dismissed out of hand. As one prestigious journal put it, any work not constructed from Newtonian principles 'is absolutely *wrong*'.[25] Yet ironically, there was no common agreement on what these basic building blocks might be. Opponents delighted in pointing out that Newtonians could not even decide among themselves whether attraction should be regarded as the *effect* of gravity, or as an inherent property of matter that *caused* gravity, the visible manifestation of God's power that could never be understood. These were crucial distinctions for Enlightenment philosophers preoccupied with the extent of God's intervention in the smooth running of the universe.

Science, religion and philosophy have now been demarcated into separate areas of specialization, but for Newton and his successors, questions about God played vital roles in how his theories were received. Influential critics objected that by making particles or planets attract each other, Newton had blurred the distinction between matter and spirit, thus minimizing God's special role in the universe. They worried that this would open the door to materialist philosophies and promote the spread of atheism, which was perceived as a major danger to the established social order. Samuel Taylor Coleridge critically commented: 'It has been asserted that Sir Isaac Newton's philosophy leads in its consequences to Atheism; perhaps not without reason, for if matter by any powers or properties *given* to it, can produce the order of the visible world, & even generate thought . . . where is the necessity of a God?'[26]

Newton's works were open to conflicting interpretations. English clergymen preferred to emphasize those aspects of Newton's work that best suited their own interests. Concerned about threats to Christianity, they tried to consolidate their own positions by focusing on those sections of Newton's writings which maintained that God is constantly present throughout the universe and can intervene in its operation. One chaplain graphically portrayed Newton's gravity as 'the immediate Fiat & Finger of God' that would help to 'undermine & ruin all the Towers & Batteries that the Atheists have raised against Heaven'.[27]

In complete contrast to this insistence that Newton's physics supported orthodox Christianity, Pierre-Simon Laplace – who dubbed himself 'the French Newton' – developed Newton's ideas to create a very different and Godless cosmological model. In Laplace's deterministic universe, mechanical particles move in predictable paths, governed by immutable, abstract laws of nature. When Napoleon accused him of excluding God from his system, Laplace retorted abruptly that he had no need of that hypothesis, a deliberate echo of Newton's phrase 'I feign no hypotheses (*hypotheses non fingo*).'

Simplified versions of the past can carry great polemical force, and it has suited the interests of historians and scientists to present the rise of Newtonianism as a singular success story. But as these examples of religious readings illustrate, being a Newtonian meant different things to different people. Countless practitioners throughout Europe came to describe themselves as Newtonian, but they held contrasting opinions on fundamental questions about the nature of the universe and the conduct of science. The Newtonian edifice was made to appear indestructible, yet it was riddled with enormous internal contradictions. These arose partly because Newton himself was far from consistent, but more significantly because his followers found it advantageous to interpret his legacy in various ways.

Since Newton's death, most historians have narrated only those features of his life that consolidate their image of an ideal researcher into the laws of nature. Although Newton is commemorated as a great scientist, men whose work overlapped with his to a great extent – for instance, Descartes, Leibniz, George Berkeley – are remembered mainly as philosophers. At the same time, Pope – Newton's Enlightenment partner – has become far less famous, but is firmly entrenched in the literary canon.

This close emphasis on Newton's scientific achievements is partly due to the great transformations in what we consider to be worthwhile knowledge, since modern science enjoys vastly greater social prominence than some of the scholarly pursuits valued by Newton and his contemporaries. Relatively few people are now interested in millenarian prophecy, ancient dynastic dating systems or techniques for milling coins, but such topics preoccupied Newton and his colleagues. The modern focus on Newton as a scientific hero owes much to the cumulative effects of 300 years of media manipulation. Newtonian propagandists felt that his alchemical inquiries and unorthodox theological inclinations threatened to tarnish his image. One way of protecting his posthumous prestige was to ensure that only works judged to be appropriate were published, and discreet monitoring processes were set in action immediately he died.

Some parts of the story have an almost farcical air. Since Newton failed to leave a will, family wrangles clouded the immediate disposal of his estate, and his huge legacy of handwritten papers was eventually inherited by the family of the Earls of Portsmouth. No systematic attempt was made to classify this archive until 1777, when a Tory bishop deliberately excluded some 'bundle[s] of foul papers' from the five-volume edition of Newton's *Complete Works* he was preparing: Newton's views on alchemy, sex and sin were not for public consumption. Fifty years later, as the antiquarian quest for ancient manuscripts boomed, scientific researchers like David Brewster assiduously searched for Newton's original documents, but selectively retrieved only those covering suitable topics. In the 1870s, a group of Cambridge scientists belatedly sifted through all the Portsmouth manuscripts, retaining those they felt to be scientifically valuable, but returning back into country-house obscurity the packets of alchemical, chronological and theological material they condemned as being 'of very little interest in themselves'.

This suppression of unfavourable evidence continued well into the twentieth century. In 1936, the Portsmouth family decided to auction off its collection, by now sadly diminished and stashed in a metal trunk. Because of an unfortunate clash with a sale of Impressionist art, the lot was grossly undervalued and sold off in portions. Its subsequent partial reassembly is due to the initiative and generosity of two private collectors who repurchased many of the scattered manuscripts from individual dealers. One of them was the biblical scholar A. S. Yahuda, a close friend of Einstein who came from the area then known as Palestine. The other was the English economist John Maynard Keynes, who later sent shock waves reverberating through scientific communities by describing Newton as an esoteric magician with one foot in the Middle Ages, a blend of Copernicus and Faustus for whom alchemy was as important as physics.[28]

Scholars have been able to access most of Newton's manuscripts since the 1970s, although they often disregard works judged to fall outside the accepted scientific canon. In contrast, the American historian Richard Westfall read virtually everything available while he was preparing the most comprehensive biography of Newton that has so far appeared, yet even he deliberately focused on what he anachronistically called Newton's 'scientific career'. In an impressive attempt at frank self-revelation, Westfall confessed to feeling dwarfed by Newton's unapproachable intellect. He stamped with academic authority Newton's retrospective canonization as the world's first scientific genius.[29]

In the past, most people were either completely unaware of Newton's alchemical and theological expertise, or else dismissed these hard-earned skills as eccentric self-indulgence. Although it is hard to overturn three centuries of effort dedicated to setting Newton up as an idealized figurehead of science, his alchemical activities are becoming increasingly recognized. Several fine historians have convincingly demonstrated the interdependence of Newton's religious, alchemical and scientific thought, and influential revisions of Newton have carried enticing titles referring to 'the pipes of Pan', 'the Hunting of the Greene Lyon', and 'the last sorcerer'.[30]

In 1936, wealthy bibliophiles had bid unenthusiastically for Newton's manuscripts. One lot of alchemical papers sold under the auctioneer's hammer has since disappeared, a lost treasure that Newtonian scholars fantasize about retrieving. At the end of the twentieth century, a team of academics was awarded a third of a million pounds to transcribe a weighty stack of other previously unpublished theological and alchemical writings. At roughly 10 pence a word, these will be catalogued for the Internet and collated into a shelf of books. For effective research, this material needs to be organized chronologically, a task made harder by Newton's parsimonious habit of writing on the back of fifty-year-old documents. As the project leader explained, such work will reveal how Newton's scientific ideas stemmed from the alchemical and theological goals that he pursued with missionary zeal. 'The psyche of the private papers showed Newton sought a more noble higher truth. The current world, he believed, was not fit to receive the truths he was about to give them.'[31]

2

ICONS

How many mansions are decorated with the portrait of the beloved sovereign – the pious divine – the sage philosopher – and the skilful physician? . . . As the absence of the sun is supplied by artificial lights, so well-finished Portraits compensate the loss sustained by the removal of the excellent originals.

John Evans, *Juvenile Pieces* (1794)

What did Isaac Newton look like? At the end of the eighteenth century, this apparently simple question was hard to answer. Images of Newton abounded. As the poet Thomas Maude wryly observed, 'Various are the effigies of Sir Isaac, both in frontispieces, medallions, busts, seals, and other engravings, but most of them are dissimilar from his monument and from each other.'[1] Jolly pottery ornaments, sombre oil paintings, brass coins, allegorical engravings, Wedgwood wall plaques, imitation marble busts – there were versions of Newton on sale to suit every taste and budget. But even a rapid glance makes it clear that Maude was right in saying how much they differed from each other. We have no definitive image of Newton.

The picture that is the most familiar nowadays was commissioned by Newton himself from Godfrey Kneller, renowned and expensive portrait painter to the King (*Figure 2.1*). Strikingly similar to Salvator Rosa's self-portrait, Newton's thin, pale face, dark clothes and unkempt natural hair conform to seventeenth-century conventions for depicting a melancholy natural philosopher or a religious anchorite. This was, perhaps, how Newton appeared when he visited Kneller's studio on a trip to London in 1689, two years after publishing his famous book on gravity, when he was still cultivating his role as a scholarly Cambridge recluse.[2]

During the eighteenth century, very few people were aware of this picture's existence. Only Newton's close friends would have seen it, on display in his own house along with several other portraits and busts – a common form of gentlemanly self-advertisement. Kneller's second portrait of Newton, painted thirteen years later, was far better known because it was widely reproduced (*Figure 2.2*). By then, Newton enjoyed a prominent metropolitan status as Master of the Mint and lived in a comfortably furnished Westminster house, cared for by his niece and several servants. Now aged sixty, Newton chose to present himself very differently. Swathed in a sumptuous red robe and wearing an elegant dark wig, this learned man of the world matched Enlightenment ideals of fashionable, well-fed sociability.

Victorian connoisseurs later sneered at this 'affected representation of Newton the dandy ... the prosperous man of the world, with a carriage and horses, and with three male and three female servants'. But Newton's contemporaries judged it quite appropriate for a scholarly gentleman to be portrayed in bright colours rather than the dark sobriety that was to become conventional in the next century. They regarded it as a sign of serious study when, instead of his tight-fitting jacket suitable for public appearances, a man wore a long flowing gown (called a banyan) that enabled his body to relax and his mind to concentrate. As Newton settled in to his new public role of metropolitan administrator, he commissioned engravings of this portrait as gifts for selected colleagues in England and abroad. In addition, miniatures and commercial reproductions ensured that during the eighteenth century this became one of the best-known images of Newton.[3]

The other popular picture of Newton during the eighteenth century was by John Vanderbank (*Figure 2.3*), and – like Kneller's – became extremely well known through its engraving, which formed the frontispiece of the definitive third edition of the *Principia*. Although Newton was by then an incontinent invalid aged eighty-three, Vanderbank portrayed him as an elegant, alert scholar dressed in crimson, Newton's favourite colour and traditionally used to denote royalty or aristocracy. Most people kept tactfully quiet about the way Vanderbank had flattered his sitter, but one American visitor condemned the Royal Society for allowing such a false image of its elderly hero to be disseminated. He reported candidly: 'by all those who have seen him of late, as I did, bending so much under the Load of Years as that with some difficulty he mounted the Stairs of the Society's Room. That Youthfull Representation will I fear be considered rather as an object of Ridicule than Respect, & much sooner raise Pity than Esteem.' But forty years later, when few eye-witnesses survived, this deceptive representation had become authenticated as being 'extremely like Sir *Isaac*, as we are informed by those who knew him many Years'.[4]

Artists and engravers frequently copied Vanderbank's frontispiece, which appeared only the year before Newton died. Pirated versions, executed with varying degrees of skill and fidelity to the original, illustrated journal articles and entries in biographical dictionaries. This meant that many people who had never even opened the *Principia* became familiar with this image, even if it did become somewhat distorted through frequent copying. An enthusiastic bookseller in Amsterdam incorporated the portrait into his trademark, while several artists purloined it for collective portraits of great men; hung in hospitals and training institutions, these were intended to inspire their viewers with visions of England's geniuses.[5]

The abundance of relatively cheap engravings, busts, statues, coins and ephemeral media meant that many people encountered Newton in the course of their normal daily activities, not merely when they chose to take a scholarly tome down from its shelf. Particularly for intellectual figures such as Newton, historians usually study texts rather than pictures, and neglect the valuable historical information that can be gleaned from material artefacts. Official portraits display how sitters and their peers chose to advertise themselves, while representations designed for wider audiences reveal how their reputation was fashioned. Even blatantly fictionalized images embody truths about the meanings of a cult figure. It was not the search for an authentic likeness, but the freedom to provide convincing visions of how Newton *might* have looked, that was to prove so significant for his elevation into the first scientific genius.

Visual materials are especially rewarding to study for this period. For one thing, people expected pictures to carry a moral message. As one artists' manual explained, 'All the fine arts have a double purpose; they are destined both to *please* and to *instruct*.'[6] In contrast with our post-Freudian era, during the eighteenth century portraits were valued not so much for exposing individual idiosyncrasies as for providing the spectator with a morally improving role model. Blemishes were tactfully disguised to leave a more flattering image for posterity.

The Bishop of Rochester, who knew Newton for twenty years, remarked to a friend that 'in the whole air of his face and make, there was nothing of that penetrating sagacity which appears in his composures'.[7] As the Bishop realized, painters emphasize different characteristics to convey a sitter's character and significance, so that visual representations of Newton depend on when they were produced, by whom and for what reason. A portrait was, of course, intended to resemble its subject, but it was also meant to represent an idealized version of a particular type of person, such as a military hero, a virtuous woman or – as in this case – a dedicated scholar. The commissioner, the sitter and the artist all influenced the picture's final appearance, to which spectators also brought their own interpretations. As a consequence of these interactive relationships, portraits simultaneously shaped and reflected ideological constructs such as national character, appropriate gender behaviour and class structures.[8]

A second reason for emphasizing the value of visual sources is the explosive expansion in the popularity of engravings and reproduction sculptures towards the end of the eighteenth century. As the art market became commercialized, cheap copies of original representations became increasingly popular to provide what was called 'decorative furniture' for people's houses. James Watt's father, an impoverished Clydeside teacher of navigation, owned only two portraits, one of the Scottish mathematician John Napier and one of Newton. Like more prestigious collectors, he probably placed these prominently not only to encourage his children, but also to display his national and intellectual allegiances.[9]

At the end of the seventeenth century, Newton was known only to a small group of natural philosophers, mainly at Cambridge. A hundred years later, he was universally celebrated. Newton's fame did, of course, spread through books, poems and lectures, but investigating the changes in his visual imagery reveals how he became a familiar icon, and entered the homes and minds of many who knew little about optics or gravity. Newton and his friends commissioned about twenty major portraits before he died, as well as several busts, engravings and plaques. Art historians and biographers focus almost exclusively on these expensive originals, but they were only seen by restricted circles. Because engraved copies and less sophisticated representations reached wider audiences, they played an important part in creating Newton's public image.

Self-fashioning

Although biographers have maintained that Newton shunned fame and was uninterested in art, the visual evidence suggests that he actively intervened in fashioning his public image. He sat for over twenty busts and portraits, paying large sums of money for several of them himself. While vanity may have influenced the deceptive youthfulness of his portraits, the extent of this self-representation was not unusual in England at this period. As the Whig journalist Richard Steele commented, 'No Nation in the World delights so much in having their own, or Friends, or Relations Pictures . . . Face-Painting is no where so well performed as in England.'[10]

Newton adopted several tactics to disseminate his appearance. Although no pithy quotations survive to indicate his preferences, we can infer his attitudes by examining his behaviour. The boundaries between home and office were less well defined than now, so that Newton's close colleagues visited him in his own house and would have seen his pictures there. Like other eighteenth-century English gentlemen, Newton displayed his own image prominently in his London house, but we have only scanty evidence of precisely where he hung his Knellers and portraits by other eminent society artists. We do know that when he was in his seventies, he sat several times to the sculptor David Le Marchand, commissioning an expensive bust and an ivory plaque that he kept in his dining room, where guests could admire the good taste that only comes with wealth.[11]

To impress wider circles of viewers, in 1717 Newton donated a large portrait by Charles Jervas to the Royal Society, so that he was also on show in a more public location. In accordance with a practice that Newton himself initiated for the Society's portraits, gold letters across the top informed admirers that Newton had been President since 1703. To reach those colleagues living outside London or abroad who never entered the Royal Society, Newton sent pictures of himself as inducements or rewards for a particular favour. The 1702 Kneller (*Figure 2.2*) is so well known partly because private engravings were commissioned – most probably by Newton, or at least with his permission – and miniature copies were painted for gifts.

Because portraits could be used to express power relationships, they played an important role in patronage networks. Exchanges were often negotiated through intermediaries. When Richard Bentley – then Master of Trinity College, Cambridge – commissioned James Thornhill's first portrait of Newton, he achieved two aims through this act of apparent generosity: he acknowledged Newton's cooperation in letting him publish a second edition of the *Principia*, a project that would benefit Bentley financially and socially; and he solicited Newton's support for his unpopular College reforms.

This portrait, which still hangs in the Master's Lodge at Trinity, shows Newton gazing out as if lost in thought and unaware of the viewer's presence. Although draped in his ornate philosophical banyan, most unusually for this period he wears no wig, but holds his hand near his heart in a traditional gesture of truth. Newton commissioned Thornhill to paint a second and rather different version, which is now in Woolsthorpe Manor, Newton's childhood home (when Newton's London house was demolished two centuries later, Thornhill's visiting card was found under the floorboards). Dressed in a different banyan, Newton chose to present himself as a visionary Roman senator. Following the contemporary gentlemanly obsession with Roman culture, a classical column has replaced the conventional background landscape of the earlier picture, and he holds his right hand in a classical oratorical gesture.[12]

It was this remodelled portrait that Newton decided to have engraved. Conducting his end of the correspondence through Bentley, he sent a copy to reward the mathematician Roger Cotes for his editorial work on the second edition of the *Principia*. In his letter of thanks, Cotes rather ungratefully 'wish'd it had been taken from the first of Mr Thornhill's', but other natural philosophers were keen to acquire this picture as a symbol of recognition. Their eagerness enabled Newton to convert portraits into assertions of power by withholding them. In 1721, when the French mathematician Pierre Varignon diplomatically conveyed Johann Bernoulli's desire for a portrait, Newton – who perceived Bernouilli as one of Leibniz's allies in the protracted international debates about who invented the calculus – refused to comply without a public acknowledgement of his priority claims going back fifty years.[13]

Concerned about the reception of his ideas in France, Newton had for many years cultivated the friendship of Varignon and other influential Parisians. However, it was only when he hoped to sway negotiations for a de luxe French edition of the *Opticks* that Newton decided to gratify Varignon's earlier request for a portrait. For this gift of international diplomacy, he again turned to Kneller, this time choosing to be depicted as an elegant bewigged gentleman, gloves in his hand and sword at his side (*Figure 2.4*). Since he was actually approaching eighty, Newton would have been delighted by Varignon's comment, written to a colleague but surely intended to reach him, that he appeared no more than fifty years old. Varignon already owned an engraving of Newton, but flatteringly wrote back that 'when the canvas had been unrolled your most high and lofty intellect together with the dignity of your aspect seemed to me as it were alive in the countenance and forehead and eyes of this likeness'. He reciprocated with a portrait of himself that Newton hung in his bedroom.[14]

Newton only allowed himself to be shown full- or half-face. In expensive oil portraits, eighteenth-century men were rarely shown in profile. This may have been for practical reasons, such as disguising an absence of teeth or preventing the face from being obscured by a bulky wig. Profile images were almost entirely restricted to far cheaper media: coins and medals near the beginning of the century, and later on, silhouettes in paper, glass or pottery.[15] Kneller encouraged his sitters to bring a colleague, so while Newton's gift to Varignon was being painted, he chatted to his friend William Stukeley, most famous for his researches into Stonehenge and Druidism. But when Stukeley suggested that Kneller should paint him in profile, Newton angrily refused, retorting, 'What! Would you make a medal of me?'[16]

Stukeley did in fact make several sketches of Newton, although only one has survived in his private papers, and was not published until 1938 (*Figure 2.5*). Since it does indeed show Newton as a Roman emperor on a medal, presumably Stukeley never dared show it to his elderly friend. Although this picture seems quite extraordinary to us, during the eighteenth century Nature was frequently represented as the multi-breasted Diana of Ephesus, an ancient symbol of fertility. The comets in the background probably refer to Newton's analysis of the 1680 comet as well as to Stukeley's own interest in the Milky Way, which he discussed with Newton. In Roman mythology, the Milky Way was formed when Juno's milk spurted across the heavens.

At that time, gentlemanly collectors purchased and exchanged engravings, coins and medals of elite 'heads', methodically arranging their subjects in books and drawers, and classifying them according to their social station (a practice later known as Grangerization, after its most famous adherent James Granger, a biographer and clergyman). Despite his own obsession with ancient chronology, Newton was renowned for despising antiquarian pursuits like coin collecting, and sneeringly remarked of an archaeologically inclined former President of the Royal Society, 'Let him have but a stone doll and he's satisfied.'[17]

After Newton died, he could no longer exert control over how he was represented. Only five years later, an allegorical engraving showed his head in profile on a large coin, thus simultaneously hinting at his regal status and celebrating his long employment at the Mint (*Figure 2.6*). Queen Caroline and other admirers arranged for high-quality commemorative medals to be struck, and most of these also showed Newton in profile, often appearing like a Roman emperor. Newton was incorporated within a collective pantheon of worthy Enlightenment Englishmen, just one among his contemporaries who, rendered as classical heads, were destined to circulate between collectors' cabinets and be illustrated in catalogues.[18] When the influential essayist Joseph Addison commented that 'A cabinet of Medals is a collection of pictures in miniature . . . You here see . . . the whole catalogue of Heroes', Pope quipped back that

> *Future ages with delight shall see*
> *How Plato's, Bacon's, Newton's looks agree.*[19]

Celebration

During the eighteenth century, foreign writers often remarked on the English innovation of honouring men for their military or intellectual achievements with ceremonies formerly reserved for royalty. Voltaire marvelled at the splendour of Newton's funeral procession, observing that he was buried like a king who had treated his subjects well. Forty years later, a Piedmontese visitor commented 'that the practice of honouring men of talent with statues is still vigorously maintained by the English, who follow the Greeks and Romans in their esteem for talent as well as in their love of liberty'.[20] In contrast with France, this programme of publicly commemorating the nation's heroes was initiated not by the state, but by private individuals pursuing their own agendas. Various projects that promoted Newton were launched with diverse objectives, and as knowledge of them spread, they collectively contributed to making Newton a British hero. Newton's image was prominently displayed in three major locations: Westminster Abbey, country estates and Cambridge University.

Newton's most influential early hagiographer was John Conduitt, who erected the splendid marble monument to him in Westminster Abbey. He had married Newton's wealthy and beautiful niece Catherine Barton, despite the rumours that it was only through her sexual liaison with the Earl of Halifax that Newton had been appointed Master of the Mint. Conduitt organized a posthumous publicity campaign that inevitably cast glory on himself by strengthening his associations with his famous relative: his own funeral plaque in Westminster Abbey even starts with a tribute to Newton.

In 1732, five years after Newton's death, Conduitt commissioned William Hogarth to paint a scene from John Dryden's fashionable play, *The Indian Emperor* (*Figure 2.7*). Ostensibly portraying a children's theatrical performance in Conduitt's town house, this conversation piece is permeated with references to Newton. Hogarth subtly made visible the force of Newtonian gravity. In the audience, a royal governess commands her own daughter to pick up her fallen fan, while the juvenile actors on the stage seem bonded together by this universal power of hidden attraction, a common metaphor for a moral and harmonious society. In the background, prompting the children but with his back turned to such light-hearted entertainment, Hogarth has slipped in Newton's experimental collaborator John Desaguliers, peering myopically into the play's text. The eminent spectators are dominated from the mantelpiece by a marble bust of Newton, while Conduitt and Barton, Newton's closest relatives, look down on the proceedings from their portraits on the wall. These emblems of science and gravity (in every sense of the word) eclipse the statues of an artistic muse and frivolous, disorderly Pan, and are placed above the three royal children in their little box. True authority, Hogarth implies, lies with Newtonian reason, the power that will enable these aristocratic children to ensure Britain's continued might.[21]

The bas-relief below the bust is a direct reference to Conduitt's magnificent monument to Newton at Westminster Abbey, completed the previous year (*Figure 2.8*). In contrast with the £20 he pledged to the poor of Newton's native parish, Conduitt spent over £700 on this spectacular public tribute. Newton's monument was one of a contrasting pair that, constructed at around the same time, separated the eastern part of the Abbey from the western nave. The Abbey was becoming increasingly important as a tourist attraction, and the nave became a public space for celebrating Enlightenment achievements. French writers commented disparagingly on the absence of coherent order among the nave's funerary tributes, which were individually designed and privately funded. Unlike the Gothic stone commemorations of royalty in the older religious centre behind the altar, classical marble monuments were chosen by families and friends to commemorate contemporary heroes. This distinction in style and function between the eastern and western parts of the Abbey was emphasized still further in the nineteenth century by a decorative screen, which nowadays frames Newton in coloured tracery.[22]

Although Conduitt specified many of the details of Newton's monument, he employed William Kent to make the initial sketches, which were subsequently realized in marble by the Flemish sculptor Michael Rysbrack. Backed by the unprecedently large pyramid, classical symbol of eternity, this imposing edifice ensured the continuity and consolidation of Newton's social persona despite the disintegration of his physical body. Newton seems to be declaiming, yet his horizontal posture and the mournful pose of Astronomy (Urania) on top of the globe suggest that he is dead. Through such ambiguities, this secular memorial conformed with Roman representations of life after death, as well as contemporary religious insistence that the virtuous dead pass into the afterlife with their works.[23]

The iconography illustrates how our assessments of Newton have changed. We celebrate him primarily for his work on gravity and light, but his elbow rests on four tomes rather than two, because at the time of his death he was also famous for his studies in theology and ancient chronology. Astronomy's globe shows the path of the 1680 comet that played a crucial role in Newton's theoretical work, but whose fame has now been eclipsed by Halley's comet, which appeared two years later. Still more unexpectedly for us, the globe's constellations indicate how Newton dated the voyage of the Argonauts. At his feet, the scroll not only pictures the solar system, which we regard as Newton's greatest area of expertise, but also displays a converging series, whose significance is now only appreciated by mathematicians. On the bas-relief, the industrious cherubs depict Newton's innovations in chemistry and coinage, two aspects of his work that are now largely forgotten.

While Newton's monument celebrated the rewards of intellectual labour, its partner on the other side of the aisle, also sculpted by Rysbrack, commemorated the military hero Lord Stanhope. This pair personified the traditional contrast of contemplative with active virtue. The gradual conversion of the country's oldest religious centre into a national shrine for secular achievement was imbued with political interests. Like Shakespeare and Handel, who joined him some years later, Newton was being embodied as a national hero who exemplified Whig ideals. Kent and Rysbrack were both protégés of Lord Burlington, and these two monuments exemplify how the Burlington circle and other Country opposition sympathizers appropriated the nave to articulate their political ideology. Stanhope, who had died over ten years earlier, bitterly condemned the corruption underpinning Robert Walpole's authoritarian regime.

Burlington's close friend Pope, who had been involved with Conduitt in the Newton project from its inception, wrote epitaphs for several of the men commemorated in the Abbey nave. Most famous now is his tribute to Newton:

> *Nature and Nature's Laws lay hid in Night.*
> *God said,* Let Newton be! *and All was* Light.

Often reprinted in books and journals, this became so well known that it could be parodied:

> *But* Pope *has his faults, so excuse a young spark,*
> *Bright* Newton's *deceas'd, and we're all in the dark.*[24]

Newton's monument was seen by a wider public than the tourists who flocked to the Abbey itself, because it was reproduced in engravings, medals and other pictures. Poems that seem painfully flamboyant even by eighteenth-century standards celebrated the completion of this scientific memorial and stirred public pride by sanctifying Newton as a national genius:

> *That Soul of Science! that unbounded Mind!*
> *That Genius which exalted Human Kind!*
> *Confest Supreme of Men! his Country's Pride!*
> *And half esteem'd an Angel – till he dy'd . . .*[25]

Accounts in journals hymned Newton's praises, explained the memorial's symbolism, and printed the epitaph inscribed below the statue in the original Latin, as well as in English for less-educated readers. Public interest was also roused by fierce debates about the monument's artistic merits, and it rapidly became an immediately recognizable national icon. Thirty years later, Hogarth incorporated it within his political caricature *The Bruiser*, converting Astronomy's globe into a Prussian millstone dangling threateningly over William Pitt's head.[26]

Several popular journals reproduced this chauvinistic tribute: 'This grand and magnificent monument, erected to real merit, is a greater honour to the nation, than to the great genius for whom it was raised; and in this light it is viewed by all Europe.' This privately funded monument, erected by Newton's family, had become a public expression of British intellectual supremacy. Conduitt would have been delighted to realize how substantially his extravagant gesture reinforced Newton's status as a national hero.[27]

Another individual who contributed to Newton's posthumous fame was Caroline of Anspach, wife of George II. This powerful and clever woman patronized an elite group of intellectual men, including Newton, and she took a keen interest in scientific issues. In particular, reflecting how natural philosophy was intimately intertwined with political affairs, the German princess became involved in a famous correspondence with metaphysician Samuel Clarke, who loyally defended Newton against Leibniz. Select visitors to Caroline's private apartments could see her Kneller portrait (a version of *Figure 2.1*), and when Newton died she commissioned the Mint's finest engraver to make a commemorative medal in gold, silver and bronze, at three prices to suit admiring collectors with different budgets.[28] After she died, Fellows of Oxford University produced a volume of commemorative poems in English and Latin. Most unusually for a queen, these often praised Caroline's scientific acumen and her familiarity with Newtonian philosophy ('her tow'ring thoughts would often stray / To distant Orbs, and range the solar way . . .').[29]

In contrast with Conduitt's obsessive promotion of Newton, the Hanoverian queen was concerned to demonstrate her allegiance to England. While the nave at Westminster Abbey came to house a varied collection of private monuments paid for by personal patrons, Caroline influenced projects specifically designed to provide a collective commemoration of national achievement. She engaged Kent to design a Hermitage for the royal gardens at Richmond. Built from stone, this pseudo-ancient retreat nestled into the side of a hill, but was later demolished by Capability Brown. From the Hermitage's rustic exterior, a tented entrance led to a formal octagonal interior with niches destined to hold the busts of five men Caroline regarded as the cream of the country's intelligentsia.

Time has eroded the reputations of some of these figures as harshly as their stone busts. Caroline had different priorities from modern biographers, and she reserved the splendid central niche decorated with rays of gold not for Newton, but for Boyle, whose glory was gradually eclipsed by Newton's during the eighteenth century.[30] Of the remaining three, Locke is still extremely famous, whereas Clarke and William Wollaston, a religious philosopher, are scarcely remembered.

Although access to the grotto itself was limited, engravings of the Hermitage and the five busts were marketed to wider audiences as sets and also reproduced in books and journals, often accompanied by eulogistic verses. Newton's role as a national intellectual exemplar was further enhanced by a poetry competition in the *Gentleman's Magazine*, and by journal articles debating Caroline's patronage of the arts. While some critics satirically compared her to Louis XIV, others praised her decision to choose British heroes like Newton for her '*Monuments of Genius*' rather than Leibniz, her compatriot but Newton's bitter rival. For one writer at least, Caroline had become the people's princess: 'When her Majesty consecrated these dead Heroes, she built herself a Temple in the Hearts of the People of England.'[31]

Caroline's patronage benefited Kent enormously, and he designed a second collective monument to national genius, the Temple of British Worthies for Lord Cobham's garden at Stowe, England's most famous landscaped estate. With their statues, temples and follies, eighteenth-century gardens were carefully landscaped to provide beautiful natural settings for walking, meditating or engaging in sexual and political intrigue.

Carrying classical as well as modern allusions, Stowe's Elysian Fields were designed to be read allegorically and, rather like the nave of Westminster Abbey, embodied a political manifesto celebrating Whig ideals and achievements.[32]

The Temple of British Worthies, a curved monument whose niches hold busts of famous national heroes, was deliberately placed facing the Temple of Ancient Virtue across a stream symbolically labelled the Styx. The sixteen worthies chosen to celebrate British achievement displayed Whig virtues that Cobham found lacking in Walpole's administration, and some of them are contemporary politicians who are no longer famous. A central bust of Mercury (classical leader of souls) separated active heroes – such as Walter Raleigh and Queen Elizabeth – from the men of thought, who included Newton, Bacon and Locke. The longest inscription and the most prominent position were reserved for Pope, who had played an influential role in the garden's design. As a natural philosopher, Newton represented the disinterested search for objective truth, but there were also close ties between Whig political aspirations and Newtonian projects of improvement.

For perceptive strollers, the Elysian Fields provided a commentary on modern corruption. Thus the classical heroes in a temple at the top of a hill were full-length statues on pedestals, while the British worthies below were only eye-level busts. In *The Seasons*, an extraordinarily popular poem couched in Newtonian imagery, James Thomson envisaged perambulating round Stowe with Pitt, conversing about the country's political degeneration. Nevertheless, the Temple of British Worthies continued to inspire sentiments of national grandeur. 'Does not your Pulse beat high,' asked the artist William Gilpin, 'while you thus stand before such an awful Assembly? Is not your Breast warmed by a Variety of grand Ideas, which this Sight must give Birth to?' But subsequent visitors became increasingly oblivious to Cobham's political intentions, and a Thomas Rowlandson cartoon of 1805 shows light-hearted picnickers enjoying the sunshine near the Temple of Ancient Virtue.[33]

In the eighteenth century alone, there were over twenty editions of guides to Stowe, so – as with the Abbey and the Hermitage – commercial publishing ventures ensured that Newton's inclusion in a prestigious set of past and present British worthies reached wide audiences. In Gilbert West's popular country estate poem, armchair tourists could imagine contemplating Newton's bust:

> *But what is he, in whom the heav'nly Mind*
> *Shines forth distinguish'd and above Mankind?*
> *This, this is* Newton; He, *who first survey'd*
> *The Plan, by which the Universe was made:*[34]

These adulatory lines now seem plodding, but they were often reproduced to cater for eighteenth-century tastes. For instance, they were printed below a second issue of the engraving shown in *Figure 2.6*, and to judge from the number of editions, the entire poem was hugely successful. It is interesting that among British heroes, West referred only to Milton and Shakespeare as geniuses. This restricted use confirms both that by mid-century scientific insight was not yet as highly valued as literary originality, and that Newton's works had not had time to satisfy the endurance criterion of genius.

At Cambridge, as at Westminster Abbey and country-house estates, individual initiative also played a vital role in consolidating Newton's posthumous fame. He was converted into one of the city's tourist attractions largely through Robert Smith, Master of Trinity College, who determinedly campaigned to enhance the College's international reputation and to establish natural philosophy as an essential component in the curriculum of a progressive Whig university. A generous donor himself, Smith solicited portraits, statues and busts of Trinity's greatest scholars, and transformed the publicly accessible buildings into galleries glorifying College achievement.[35]

While visitors and residents could admire Newton's images at first hand, engravings and books ensured that these icons of Cambridge's intellectual splendour spread across the country. For example, the rich republican Thomas Hollis donated an oil painting of Newton, derived from Vanderbank's (*Figure 2.3*), which became a common source for encyclopaedia articles because he also published it as a fine mezzotint. Embellished below by a text from Voltaire extolling English natural philosophy, and a small cap of liberty advertising Hollis's political ideals, this engraving became so popular that collectors cut it out of Hollis's *Memoirs* to frame and hang on their walls (thus creating a frustrating absence for modern researchers!).[36]

Smith commissioned Louis-François Roubiliac for the marble statue that still stands in the antechapel (*Figure 2.9*). The relatively recent backdrop of College members who died in the World Wars alters its impact, yet confirms Newton's status as both a local and a national hero. It was this statue that later inspired William Wordsworth's famous evocation of 'a Mind / Voyaging thro' strange seas of Thought, alone'. Roubiliac's Newton is an unusually early example of a secular commemorative statue erected in a public location. In contrast with Kneller's portraits or Rysbrack's monument, Roubiliac sculpted Newton neither as a reclusive scholar nor as a classical philosopher, but as a young and forceful public figure personifying male beauty and Enlightenment rationality. He paid close attention to Newton's elegant clothes, emphasizing their fineness by the creasing shown in the marble, and carefully adding seams to his shoes after overhearing a cobbler's loudly whispered criticism.

Newton appears to be lecturing, tapping his prism – held like an emblematic baton of power – in a traditional gesture of disputation. His upturned eyes and slightly open mouth closely match the illustration of ecstasy by Charles Le Brun, the Versailles designer whose treatise on depicting the passions influenced artists throughout Europe. Whereas earlier representations of great men conventionally showed them pointing to their achievements, Roubiliac has depicted a genius enraptured by divine inspiration.[37]

Smith also refurbished the College's Wren Library, enhancing Newton's fame as he promoted the academic science taught at

Trinity. Decorating libraries with sculpted busts rather than conventional painted portraits was becoming increasingly fashionable, and Smith persuaded former students to donate busts of natural philosophers and other eminent men associated with Trinity. Among the first to arrive was a Roubiliac bust of Newton, carefully placed by Smith in a prominent position across the aisle from Francis Bacon, the Royal Society's major ideological figurehead.

Smith's innovative strategy enabled him to advertise the continuity of Trinity's splendid past with the future rewards promised by scientific research. In a letter praising Smith's initiative, the blue-stocking literary hostess Elizabeth Montagu articulated the importance of material objects in reinforcing the growing split between what we now call the two cultures: 'so fine a Temple of the Muses should be adorned with all the arts of the ingenious as well as the studious nine [muses] especially in an age that honors the polite arts more than severe science'.[38]

The Wren Library was, however, a collective tribute: Newton was still not necessarily singled out as an exceptional man. In an essay on genius of 1755, William Sharpe commented that Cambridge was an ideal environment for fostering the development of innate genius, but – strangely for us – he drew no distinction between his three examples, Nicholas Saunderson, William Whiston and Newton. The first two mathematicians were also famous in their own lifetimes, but now we celebrate only the third.[39]

Through his will, Smith financed a large and colourful glass window that, completed in 1774, still dominates the Wren Library, although the design was hugely controversial and at one stage was concealed from tourists by a curtain. Clad in vibrant yellow, the muse of the College anachronistically presents its two most eminent natural philosophers, Bacon and Newton, to George III: three centuries of achievement are melded into a single image. Newton himself, dressed in blue medieval robes, is here a young man rather than a venerable sage. To modern eyes the window gives a church-like aura to this hallowed scholarly retreat, as if these two academic founding fathers have been converted into secular scientific saints.

As well as celebrating Trinity's prowess, the scene promotes natural philosophy as a nationally beneficial enterprise. Bacon had influentially stressed the practical value of publicly funded research, and he is depicted here as a public administrator dressed in his Lord Chancellor's robes. The financial rewards promised by research are symbolized still further by the hovering presence of Mercury, the god of commerce. The bright red, white and blue on the shield to the King's left, and the Britannia-like Minerva (goddess of wisdom) behind his throne, emphasize the national importance of Trinity's great men. As Newton points to his globe, the traditional symbol of a natural philosopher, he offers the fruits of his studies to the British nation and receives a victor's wreath appropriate for a national hero.

Commercialization

An eighteenth-century print entitled 'Truth sought by philosophers' shows 'PHILOSOPHY reprysented by a stately Woman' conducting a long line of male scholars towards the naked female figure of Truth. In this ornate allegorical image, Newton heads a line of famous Greek philosophers such as Socrates, Plato and Aristotle. The extended caption explains that Newton is unmistakable both because of his position as leader and by the books he holds in his hand. However, a virtually identical French engraving, originally intended to be the frontispiece of a new biography of Descartes, makes it apparent that a plagiarist had simply substituted Newton's name wherever Descartes's appeared. (This picture had caused a minor scandal in France in 1707, when reactionary scholastics who opposed the overthrow of Aristotelian ideas interpreted Philosophy as Queen Christina of Sweden, who had been Descartes's pupil until the northern climate proved fatal for him.)[40]

This is a particularly blatant example of how Newtonian images became marketable items. As the economy boomed during the eighteenth century, the range and number of goods for sale grew at such an explosive rate that the country is often said to have undergone a commercial revolution. The classic study illustrating the birth of English consumerism describes how, in the last third of the eighteenth century, the potter Josiah Wedgwood constantly introduced new products and developed sophisticated marketing techniques for advertising his wares to an expanding pool of purchasers.[41] Natural philosophy was becoming increasingly fashionable among the polite classes with disposable incomes, so Wedgwood opportunistically produced plaques and busts of famous men like Newton. Since he operated not only by satisfying demand but also by creating new market openings, Wedgwood's publicity material and products helped to boost the status of natural philosophy.

The success of Wedgwood's venture depended on the activities of previous entrepreneurs whose books, instruments and entertaining lectures had helped to make natural philosophy an essential component of an educated person's knowledge. Newton's conversion into a British scientific hero was inextricably bound up with the establishment of science as a valuable public enterprise, and with the growth of nationalist feelings in an increasingly commercialized society. By the end of the century, countless ephemeral items displaying England's new intellectual hero were available for purchase. Individuals such as John Conduitt, Queen Caroline and Robert Smith had created visual representations of Newton that helped to effect his entry into the pantheon of great Englishmen. Subsequently, enterprising artists manufactured and marketed an expanding stream of engravings, busts, plaques and medals, so that physical images of Newton from Westminster Abbey, country estates and Trinity College reached still wider audiences. These commercial products played a vital role in consolidating Newton's reputation, but their importance has so far been overlooked.

The most spectacular case of Newtonian commercialization

was the commemorative painting commissioned by an Irish entrepreneur, Owen McSwiny, with the explicit intention of marketing derivative engravings (*Figure 2.10*). Recalling religious imagery, a divine ray of light shines over Newton's funeral urn to be refracted by a prism. Against the subdued browns of the classical background architecture, the spectral colours are repeated in the bright robes of the scientific muses weeping near the picture's foot, and of the ancient and modern philosophers studying diagrams and instruments. Painted by Giovanni Pittoni and the Valeriani brothers, this huge picture was one of a series McSwiny coordinated in Italy. McSwiny planned twenty-four elaborate allegorical images of famous British men including Locke, Boyle and the Duke of Marlborough.

To recoup his expenses, McSwiny sold the originals separately to wealthy customers for their private mansions, but marketed sets of engravings to a broader, less wealthy, public. Operating at an international level, he disseminated an image of Newton as member of a distinguished elite, thus providing a secular, patriotic equivalent of older religious iconography that portrayed new national saints of the Enlightenment. McSwiny advertised them as '*British* Worthies . . . who contributed largely to carrying the Reputation and Credit of the *British* Nation to a much higher Degree than it was ever at before', proclaiming that his 'Collection of Pictures [was] fit for the Gallery of a Man of Taste . . . that it might come into the hand of Many'.[42] He thus articulated how the elevation of Newton and some of his contemporaries into civic heroes was embedded within the broader historical processes of nationalization and commercialization in a polite society.

Art historians are fascinated by McSwiny's complex and ornate scenes, but, perpetuating the elitist distinction between 'high' and 'low' art, they have ignored less ostentatious commercial ventures that seem iconographically less interesting. These were, however, significant because they collectively affected public perceptions of a heroic Newton. George Bickham's allegorical engraving was probably seen by far more people than McSwiny's, since it was published twice and a copy hung in a church near the Mint (*Figure 2.6*).[43] It first appeared five years after Newton's death, but half a century later, when Newton was a cult figure and the market for cheap engravings had expanded, Bickham's son republished it with eulogistic lines taken from a Stowe country-estate poem. Following Renaissance tradition, Newton – the 'Prince of Philosophers' – has been posthumously eternalized as a star (although the cherub at the top appears to be gazing at the goddess of mathematics through the wrong end of his telescope).

Like the Bickhams, as Newton's reputation became established, people utilized his fame to promote themselves. These self-reinforcing processes thus also consolidated Newton's heroic status. For instance, when the ivory sculptor Le Marchand had become eminent enough to have his own portrait painted, he chose to present himself with the bust he had carved of his most famous client – Newton.

At a less prestigious level, seven years after Newton died, the architectural designer Henry Bridges commissioned an engraving to advertise his sensational 'Microcosm, or The World in Miniature, Built in the Form of a Roman Temple'. Resembling a three-tiered wedding cake, this intricate astronomical clock was one of the most elaborate examples of Georgian mechanical virtuosity. With 1,200 internal wheels and pinions, the gigantic decorated machine played music while mechanical birds soared in flight and miniature workmen laboured. As Bridges toured round Britain and America, he liberally distributed his publicity poster, which was dedicated to Lord Chandos, an influential wealthy patron of Newtonian engineering projects. Many people must have admired this drawing. Following the conventions of Roman iconography, the clock is flanked by two medallion portraits. One shows Newton, copied from the frontispiece of the *Principia* (*Figure 2.4*). Bridges himself was the other subject, thus deceptively implying that Newton had endorsed his enterprise of philosophical showmanship.[44]

As the markets for pictures and *objets d'art* expanded, entrepreneurial artists retailed increasingly cheap versions of elite representations to ever-widening markets. For much of the century, the most important source was Kneller's 1702 portrait (*Figure 2.2*). Although during Newton's lifetime only select audiences could view the painted replicas and miniatures of this image, Conduitt subsequently allowed his inherited original to be engraved for Thomas Birch, a self-educated historian who later compiled the eighteenth century's most important biographical dictionary (the major printed source of personal details about Newton's life) (*Figure 2.11*). Birch's prestigious set included over a hundred 'Illustrious Persons of Great Britain': along with colleagues such as Pope, Boyle and Locke, Newton was welcomed into a British pantheon that included Geoffrey Chaucer, Henry VIII and Francis Drake (virtually the only woman who ever appeared in this august male company was Elizabeth I).

All the superb engravings were adorned with allegorical decorations (called 'proper Ornaments' in the newspaper advertisements). Some of these were executed by a young apprentice called Thomas Gainsborough, although there is no record of whether he worked on Newton's picture. Among Newton's attributes were a traditional lamp of truth, a mathematical muse, and leaves of aloe, the plant that – like Newton's genius – flowers only once in a hundred years. In the background, the pyramid represents eternity and recalls the Westminster Abbey monument. Like those of his 'Illustrious' companions, Newton's engraving was initially sold separately to be glazed and hung on the walls of a gentleman's study; this ambitious project proved so successful that two editions of the bound set were published.[45]

Less privileged purchasers could see a pirated and much-simplified version in the *Universal Magazine*, reversed in direction because the engraver had traced it. The accompanying hagiographic article explained how Newton's fame had spread abroad, relating the anecdote that Conduitt helped to publicize of the aristocratic French mathematician who inquired whether Newton ate, drank or slept like other men, allegedly declaring that he was a heavenly spirit rather than a human being.[46]

Like other myths of genius, this celebrated tale exemplifies the new canonization of secular intellectual geniuses, as Newton inherited the superhuman capacities of survival previously attributed to saints. Whereas Birch's skilled engraver had faithfully replicated Kneller's portrait, as the demand for cheap prints grew, less talented copyists marketed derivative versions that became decreasingly like the original Kneller (*Figure 2.12*). These second-rate imitations achieved a sort of Newtonian brand identification in the journals, encyclopaedias and collective biographies of the late eighteenth and early nineteenth centuries, both in Britain and abroad. They were sometimes accompanied by a careful copy of Newton's signature, reflecting the contemporary vogue for autograph collection (an obsession cursed by historians now faced with the task of identifying correspondents whose signature has been snipped off the bottom of letters).[47]

As cheap educational and biographical publications proliferated, engravings like these provided an important route for making Newton's face a familiar image. In addition to seeing Newton's paper portraits hanging on walls or reproduced in books, people could also buy more substantial versions of their hero. One optician living in fashionable Richmond Hill embellished his garden with a prominent gilded bust of Newton, prompting one of his neighbours to write a long yet tongue-in-cheek poem about this urban shrine to a 'demigod'.[48]

Busts are now an unfashionable genre of portraiture, but during the eighteenth century, they became indispensable decorations for libraries in private homes as well as public institutions such as schools and hospitals. A French tourist noted in wonder how English aristocratic palaces, public meeting places and gentlemanly homes were all 'adorned with figures painted and engraved, and with busts of all sizes, made of all sorts of materials, of Bacon, Shakespeare, Milton, Locke, Addison, Newton'.[49] Reproduced in varying qualities, Newton came to feature in houses, lecture halls and meeting rooms throughout the country. As Beau Nash converted Bath into a fashionable tourist resort, he commented wryly on the placing of his own portrait between busts of Newton and Pope in the Long Room. Samuel

Taylor Coleridge described an assembly of British intellectual heroes in the Philosophical Institution near Fleet Street: 'A spacious handsome room with an academical Stair-case & the Lecture room itself fitted up in a very grave authentic poetico-philosophical Style with Busts of Newton, Milton, Shakespeare, Pope & Locke behind the Lecturer's Cathedra'.[50]

Especially in the first half of the century, Newton was far more often included in such a national pantheon than displayed on his own. In particular, although Pope is less famous now, far more portraits and busts were produced of him than of Newton. William Hayley articulated the century's preference in a comic verse written for his fellow poet Anna Seward when he discovered that his London stone-cutter had sent him a bust of Newton instead of Pope:

Oh Jove! he exclaim'd if I wielded thy Thunder
I'd frighten the Sculptor who ruins my hope;
Sure never did Artist commit such a Blunder
He has sent me a Newton instead of a Pope.[51]

People with a special interest in natural philosophy chose images of Newton for their collections. When the wealthy Northumberland landowner John Bacon had his London house redecorated in the early 1740s, he placed four grisaille medallion portraits on the wall: Milton, Bacon (no relation), Pope and Newton. He commissioned Arthur Devis to paint a conversation piece, in which Bacon advertised his learning through these classicized profiles of England's intellectual heroes and the expensive air pump, telescope and other instruments displayed like costly ornaments.[52]

Similarly, for their richer customers, Rysbrack and Roubiliac used expensive marble to produce versions of Newton's bust resembling those they had created for public spaces such as the Wren Library. James West, the President of the Royal Society, commissioned Rysbrack to sculpt marble busts of Shakespeare and Newton for his new country mansion near Stratford-upon-Avon, thus linking England's traditional literary genius with the man who would become celebrated as her greatest scientific genius. The socle supporting Newton's head is inscribed with an elegiac couplet based on Ovid. Engraved in Latin, these lines would have been familiar to every educated gentleman, and can be loosely translated as 'He wanted to have a sneak preview of the heaven that he had been promised, and not go as a fresh guest to a strange house.'[53]

On the face of it, this classical quotation seems an ideal choice for commemorating Newton's work on celestial motion, an appropriate verbal equivalent of the solar system shown in Bickham's engraving, the Abbey Monument and Queen Caroline's medal. However, it was also selected for a bust of Pope, because it refers directly to one of his most famous poems, the *Essay on Man*. The identity of the quotations on these two busts underlines how in the middle decades of the century, Newton was still bracketed with other Enlightenment figureheads as a marketable commodity, rather than being singled out as a unique genius.

Rysbrack and Roubiliac charged around £40 for their marble busts of Newton, roughly the annual income of a labourer's family. But when times were hard, they sold models in painted terracotta for about half that price, as well as fragile plaster versions for a few pounds. When Roubiliac was feeling particularly pressurized by his creditors, he resorted to producing medals of Pope, Newton and Handel.[54] Capitalizing on Newton's growing fame, some sculptors (such as Joseph Nollekens) sold marble copies of Roubiliac's bust at Trinity, while other entrepreneurs started to produce even cheaper imitations. Thus John Cheere, younger brother of the famous funerary sculptor Henry Cheere, sold mock-Roubiliac busts for about two guineas.

Mocked by Hogarth for his artistic pretensions, Cheere marketed a range of statuettes to service the burgeoning demand. His Newton holds one of the classical poses traditionally favoured by Enlightenment artists, supplemented by the muse of astronomy copied from the Abbey monument. Like Wedgwood, Cheere developed special techniques to make his cheap and brittle plaster copies of Newton and other heroes resemble the robust expensive originals. He painted the bases to resemble marble, while the statues themselves were bronzed. Despite their fragility, a collection decorating a Yorkshire hospital has survived.[55]

When Wedgwood developed a new bronzing method for plaster busts, he astutely – and correctly – predicted to his business partner that 'Middling people will buy quantities at the much reduced price.' The Etruria factory mass-produced over ninety reproduction heroes, buying many of the plasters in bulk from Cheere. Evidently placing Newton in the second rank of marketable personalities, Wedgwood started with the King and classical figures, only later producing replicas of Roubiliac's bust and Le Marchand's ivory plaque.

After other portrait medallions proved successful, Wedgwood suggested making 'a suit of eminent moderns, naturalists, amateurs, etc. [to] form a constellation, as it were, to attract the notice of the great, and illuminate every palace in Europe'. In his series of 'Illustrious Moderns', Wedgwood chose his characteristic blue and white jasper for a 30-cm oval medallion of Newton with a small comet behind his shoulder. Although we now identify Newton by his apple, during the eighteenth century his comet was his defining attribute. Wedgwood was an aggressive international retailer who labelled himself 'Vase Maker General to the Universe'. Thanks to his marketing initiative, Newton transcended the bounds of a purely British pantheon to join an international assembly of over 200 eminent Enlightenment men of letters drawn from all over Europe.[56]

Wedgwood's Scottish rival, James Tassie, also contributed to Newton's fame by marketing cheap cameos. Made from a special glass he had invented, these were popular copies of the gems and rings bought and exchanged by gentlemanly collectors. Earlier in the century, the astronomer Nevil Maskelyne and the physician Richard Mead had both acquired rock-crystal seals engraved – as was the fashion – with a classic profile of Newton on one face, and intricate initials on another. Catherine the Great of Russia, an avid collector of these miniature portraits, bought an original Newton ring as well as a Tassie copy. The gift of an engraved

'GEM of NEWTON' inspired one lucky recipient to write a long patriotic poem for *The Times*. Tassie's replicas were hugely popular, and his catalogue lists hundreds of different examples. By 1795, he was advertising thirteen cameos of Newton, and at the University of Glasgow the annual prize for natural philosophy was a purse of gold and a Tassie medallion of Newton.[57]

Around 2.5 cm high, Tassie's glass Newtons cost a pound or two each, and – like Wedgwood and Cheere's busts and statues – were designed to attract middle-class purchasers. As Newton became more famous, still cheaper Newtonian mementoes came on to the market towards the end of the century. For only a couple of shillings, poorer people could buy cheerful Staffordshire statuettes of their favourite British heroes – including Newton. These pottery Chaucers, Miltons and Newtons held traditional poses, but were painted in bright colours. Holding a telescope in one hand, the youthful Newton is symbolized by a streaking comet as he unveils the globe.[58]

For an ex-Master of the Mint, the most demeaning commercial products were tiny coins minted as local tokens of exchange. Carrying crudely drawn heads of Newton, these flimsy halfpennies and farthings (quarter of an old penny) were very different from the commemorative medals struck earlier in the century. Often supplied by forgers plugging gaps in series for wealthy collectors, these mass-produced coins drew only a perfunctory visual discrimination between Newton and other Englishmen converted into classical heroes.[59]

By the end of the eighteenth century, Newton had been transformed into a national hero and scientific genius, his image at the very centre of British culture. Chauvinists regarded him as God's unique gift to a favoured nation, a sentiment encapsulated by the inscription beneath his bust in the Temple of British Worthies at Stowe:

SIR ISAAC NEWTON
WHOM
THE GOD OF NATURE MADE TO COMPREHEND HIS WORK

3

DISCIPLES

But there has been little or no publicity given to the efforts that Newton made to decode the Bible. He believed that every man owed it to God, as a debt, to use his best efforts to understand both the physical universe (the Cosmos) and his spiritual universe, the Holy Scriptures. Did you know that Newton spent a large part of his life wrestling with the symbolic code of the Bible [and] died trying to work out the date of Armageddon from it? His final calculations, performed in the early eighteenth century put the date at 1948 . . . We know today that Newton was very rarely wrong. His thought processes have been proven over time to be second to none.

Advertisement in the *Guardian*, 24 March 2001

Newton chose the title of his first book with care. Aiming to challenge Descartes's *Principles of Philosophy*, published forty years earlier, he called his own major work the *Mathematical Principles of Natural Philosophy*, thus emphasizing the importance he attached to mathematics in supplanting his predecessor. Natural philosophers learned about their universe by reading God's two great books, the Bible and the Book of Nature. Some of them felt that God had written the natural world with a mathematical alphabet, although others – the chemist Robert Boyle, for instance – thought that the language of mathematics was inappropriate for natural philosophers. Newton made his own opinion clear by closely modelling the style and structure of his *Principia* on Euclid's *Elements*, the great classical work on geometry.[1]

As Newton gained prestige, opportunistic authors started to borrow his title. Other *Mathematical Principles* have covered topics as diverse as geography, Greek architecture, biology, and spiritual metaphysics (but the *Mathematical Principles of the Theory of Wealth* arrived too late to help Newton, who allegedly lost thousands of pounds in the South Sea Bubble bonanza). The most famous instance of this genre is Alfred North Whitehead and Bertrand Russell's *Principia Mathematica* – a strategically Latin title for a book in English.

Many critics disapproved of Newton's mathematical approach. 'God forbid', scoffed Blake, 'that Truth should be Confined to Mathematical Demonstration.'[2] The first imitation *Principia* would have incensed Blake enormously. Published in 1699, John Craige's Latin pamphlet appeared in English as *Mathematical Principles of Theology: Or, the Existence of God Geometrically Demonstrated*. Proving God geometrically might seem a laughable project, but Craige was far from being an isolated crank. This distinguished mathematician was on good terms with Newton and later received an official Church post, and his work was seriously discussed in England and abroad. His book illustrates how distinctions between science and religion that seem so familiar to us did not exist for Newton and his contemporaries, who searched for harmonious laws describing how God governed both the physical and the social worlds.

At the beginning of the *Principia*, Newton's first law of motion had established the mechanical inertia of material objects like planets and billiard balls: 'Every body continues in its state of rest or of uniform motion in a right [straight] line, unless it is compelled to change that state by forces impressed upon it.' Craige transposed this terse formulation to provide an empathetic axiom for human behaviour: 'Every man strives to produce pleasure in his mind, to increase or continue in his state of pleasure.' Like Newton and their colleagues, Craige used the same intellectual tools to tackle problems in natural philosophy as in theology. One of these scholars' major concerns was to predict the second coming of Christ and the establishment on earth of a millennial paradise. Newton and Craige scoured the Bible for clues. From a text in St Luke's Gospel, Craige deduced that there would be at least one believer to welcome Christ on his return, and he set about using Newtonian principles to calculate how rapidly the strength of evidence gets diluted as it is transmitted down through the generations. He concluded that by the year 3150, the likelihood of any one person still being convinced of Christ's existence would have diminished to zero, so that Christ must (according to his logic) have appeared before that time.[3]

However bizarre it might seem now, Craige's adaptation of Newton's physics to depict human behaviour was neither a joke nor a facile comparison. Many Enlightenment writers sought laws that would portray biological and moral behaviour with the same succinctness as Newton's regulation of the physical world. The concept of gravitational attraction was applied not only to the physical world, but also to describe how living organisms function and how social groups are bonded together. The Scottish philosopher David Hume, who set out to be the Newton of the human mind, spoke for many when he insisted that 'in the production and conduct of the passions, there is a certain regular mechanism, which is susceptible of as accurate a disquisition as the laws of motion, optics, hydrostatics, or any other part of natural philosophy'.[4]

The initial reaction to Newton's *Principia* was mainly stunned incomprehension. Newton often gave the impression that he cared little about whether people understood his new ideas. Only a decade earlier, he wrote that he had 'long since determined to concern my self no further about ye promotion of Philosophy'. Behind his back, students jeered, 'There goes the man who has writt a book that neither he nor any one else understands,' while their elders confessed themselves to be equally stumped. A retired Cambridge don complained to Newton himself that 'You masters doe not consider ye infirmities of your readers, except you intended to write only to professours or intended to have your books lie, moulding in libraries.'[5]

But outside the universities, some experts objected. Knowledge, they insisted, should not be restricted to learned scholars perusing Latin texts on abstruse topics. One of these propagandists was Humphrey Ditton, a gifted mathematics teacher who was ineligible for a university education because of his nonconforming religious principles. Although he extolled the *Principia* as a 'Divine Book', Ditton scathingly compared exclusive natural philosophers with those Roman Catholics who resisted vernacular translations of the Bible. Deliberately writing in plain English for 'a Multitude of very capable Minds', Ditton protested that 'some People argue for keeping the Sacred Books in an unknown Tongue: But we pretend to a Protestant Liberty, at least with respect to our Philosophy.' His simplified account encouraged less privileged readers to discover the new cosmology and give 'thanks to the great Genius to whom all this was owing'.[6]

Ditton (duly rewarded by Newton with a job at a prestigious London college) was just one of the countless didactic authors who made Newton accessible to a wide range of readers. Privileged men might learn about Newton's theories and experiments in university courses, but many people acquired information from books like Ditton's, or by watching the performances of lecturers who travelled round the country performing experiments. Newton's ideas also moved along routes that seem counter-intuitive to us. Genres regarded as alien to modern science – poetry, treatises on aesthetics, religious tracts, paintings – were steeped in Newtonian imagery, and made vital contributions to establishing Newton's fame.

The *Principia* has come to acquire scriptural authority in modern science, but has also affected ways of thinking throughout society. Few cultural spheres remain untouched by Newton's influence. By the end of the eighteenth century, various versions of Newtonian principles were being applied in fields as diverse as politics, sociology, aesthetics and biology. At their heart lay the fundamental quest for guiding laws. Just as Newton had provided simple mathematical relationships governing the natural world, so too, it was believed, could laws be found to describe every aspect of life. When Thomas Jefferson drafted the opening sentence of the Declaration of Independence, he deliberately imbued the nation's definition of equality with a Newtonian resonance by invoking 'the Laws of Nature and of Nature's God'.[7] Saying that we live in a Newtonian universe now refers not only to how we perceive the physical world, but also to ways in which we think about ourselves and structure our society.

Books and readers

In 1693 a Latin oration was delivered in Oxford's Sheldonian Theatre by a young student, Joseph Addison. Twenty years later, he would become famous as an essayist and author of Mr Spectator's sardonic reflections on metropolitan life. Railing against the University's Aristotelian conservatism, Addison exhorted his scholarly (and probably elderly) audience to consign their books to 'Chains and Libraries, Food only for Moths and Worms' and replace them with the exciting new works on natural philosophy by 'the *great Ornament* of the present Age'. Although this was six years after the publication of the *Principia*, Addison was still talking about Descartes. But by 1737, it had become unfashionable to imply that Newton had any competitors worthy of serious consideration: when his speech was published in English, the translator informed readers that Addison had been referring to Newton.[8]

In retrospect, it seems as if a giant publicity machine was set in operation around the end of the seventeenth century to convert Newton into a secular leader whose apostles diligently interpreted his scriptures and propagated his ideas. But at the time, no central coordinator was responsible for supervising the efforts of Newton's protagonists, who were individuals concerned with promoting their own immediate interests. Describing themselves in terms that resonate with religious overtones, these Newtonian disciples campaigned against what they called the Cartesian and Peripatetic (Aristotelian) sects, seeking converts in order to boost their own reputations or to carve out a living. Although their interpretations of Newton's works diverged enormously, collectively they ensured that he became the world's most famous natural philosopher. At the same time as altering how Newton was viewed, they transformed Georgian society, starting to re-organize the boundaries between different realms of knowledge and establish new distinctions between the arts and the sciences that we now take for granted.

The Newtonian publishing industry was launched even before he died, although it was structured very differently from the modern one. Entrepreneurial authors cautiously sounded out the potential market by soliciting subscriptions in advance, cancelling unprofitable projects before they proved too disastrously ruinous. Henry Pemberton, for instance, was an ambitious young physician who managed to ingratiate himself with Newton sufficiently to be invited to edit the third edition of the *Principia*. Pipped to the post in his bid to provide the book's first English translation (*Figure 1.5*), Pemberton had more success with his next attempt to profit from Newton's growing fame. Encouraged by hundreds of positive responses to his pamphlet requesting financial support – including Newton's own advance order for twelve copies – in 1728 he published a simplified exposition of Newtonian principles, strategically including sixteen pages of subscribers' names to broadcast their enlightened generosity and advertise the book's worthiness to hesitant purchasers.

Recognizing that Newton himself was a strong selling point, Pemberton exaggerated the closeness of his own association with 'that great genius', and included a long eulogy on Newton by his sixteen-year-old friend Richard Glover. To modern eyes, Glover's lines (almost 500 of them) seem as inaccessible as Pemberton's convoluted prose, but he was catering for Georgian classical tastes. This poem was often republished separately and contributed to fashioning Newton's image. The opening lines convey its style:

To Newton's genius, and immortal fame
Th'adventurous muse with trembling pinion soars.[9]

Declaring that 'NEWTON demands the muse', Glover typified those hopeful muses who, with varying degrees of ability, hymned Newton's glory and versified his natural philosophy. These poets boosted Newton's personal reputation as well as providing basic scientific education in a palatable form.

Combining the *Principia* with the *Opticks*, Pemberton's book was one of the earliest versions of what we would now call a Newtonian scientific textbook, even though it was introduced by a long poem, was decorated with ornate woodcuts, and was devoid of mathematical formulae. Newtonian instructors worried that too much mathematics could lead to 'an overweening and most ridiculous self-conceit'. Mathematical physics was not yet being taught even at Cambridge, which was effectively a theological training centre. Most of the students were destined to be clergymen, for whom the major point of learning about Newton was that '*the knowledge of nature will ever be the firmest bulwark against Atheism*, and consequently the surest foundation of true religion'.[10]

Pemberton was targeting wealthy youths who were more accustomed to complete their education on a Grand Tour to Italy than in an experimental demonstration room. On the title-page of his *View of Sir Isaac Newton's Philosophy*, his 'view' in the vignette shows a classical temple perched picturesquely on a cliff-top, framed by mathematical instruments and a telescope. Pemberton suggested that his leisured gentlemanly readers could contemplate Newton's ideas, like beautiful buildings, 'without engaging in the minute and tedious Calculations necessary to their Production'.[11]

Other authors focused on making Newton's mathematics more accessible. In 1736, when John Colson was trying (successfully, as it turned out) to establish his suitability for the Lucasian Chair at Cambridge, he provided 'an *English* Dress' for one of Newton's unfinished books in Latin. Reflecting years of teaching experience, Colson's frontispiece attracted readers by showing Newtonian mathematics being applied to a practical problem that was close to the hearts of his gentlemanly readers: how to kill two ducks with a single bullet (*Figure 3.1*). This picture also carries subtler advertisements for Newton. Colson's superimposed geometrical diagram reinforces Newton's superiority over Leibniz by demonstrating that his mathematical concepts had physical equivalents. On the left, the Greek philosophers are using Newton's notation for their calculations in order to emphasize his continuity with classical traditions.[12]

For well-bred young gentlemen, familiarity with Newton was being converted into the academic equivalent of admiring monuments or shooting game. Women, on the other hand, were being schooled into less ambitious Newtonian pursuits. They were being provided with greatly simplified books with titles like *Newton for the Ladies*. (No doubt these were secretly borrowed by brothers and husbands unwilling to admit that they were more proficient at shooting with guns than solving equations.) Some superficial knowledge of Newton's ideas became essential for fashionable women, but they were not expected to grasp even the rudiments of the mathematics, let alone engage in serious research. Women themselves presented scientific study as a sort of moral therapy. Without learning, women become 'loitering, lolloping, idle Creatures', wrote Eliza Haywood in the *Female Spectator*. 'But of all Kinds of Learning the Study of Philosophy is certainly the most pleasant and profitable: – It corrects all the vicious Humours of the Mind, and inspires the Noblest Virtues . . . [T]he more we arrive at a Proficiency in it, the more happy and the more worthy we are.'[13]

A few women were, of course, admired for their sharp intellect. Newton himself was so impressed by Elizabeth Tollet, whom he had met in her father's home in the Tower of London, that he encouraged her education, and she became a gifted poet, fluent in several languages. Judging from her wistful comparison of her own cloistered studies with the freedom enjoyed by her brother at Cambridge, she felt almost as incarcerated as the Tower's real prisoners. In a forceful expression of her frustration, she denounced the chauvinist attitudes that kept women submissive through their scientific illiteracy:

What cruel Laws depress the female Kind
To humble Cares and servile Tasks confin'd? . . .
That haughty Man, unrival'd and alone,
May boast the World of Science all his own:
As barb'rous Tyrants, to secure their Sway,
Conclude that Ignorance will best obey.[14]

Another unusual young woman, the multilingual scholar Elizabeth Carter, translated a small Italian book into *Sir Isaac Newton's Philosophy Explain'd for the Use of the Ladies*. Belying her own impressive learning, Carter faithfully transcribed Francesco Algarotti's patronizing explanations that he had 'endeavoured to set Truth . . . in a pleasing Light, and to render it agreeable to that Sex, which had rather *perceive than understand*'. Deliberately excluding diagrams and abstract arguments, Algarotti presented Newton's ideas in a flirtatious dialogue between a flighty aristocratic woman and her condescending male tutor. Writing in the first person, Algarotti chose the narrative device of describing how he converted his Marchioness from one philosophical sect to another, but sexual innuendoes haunt his flippant references to philosophical passion, mutual attraction and the optics of mirrors.[15]

Like Algarotti, other authors also reformulated the educational tradition of simulating conversations between a naive student and a wise teacher, always casting the woman in the subordinate role, such as the ignorant younger sister of an urbane and knowledgeable brother. These contrived discussions may have imparted the basic principles of Newtonian natural philosophy, but they also reinforced women's intellectual inferiority and domestic captivity.

Although these chatty little books were explicitly written for women, to be a Newtonian lady was an impossibility, an inherent contradiction in terms. Writer after writer explained that of all the branches of learning, natural philosophy was the most unsuitable for women because it was based on rationality. Merely suggesting that a woman might have a copy of the *Principia* in her room was guaranteed to raise a laugh. Only men were judged capable of understanding Newton's ideas, although it seems that Georgian *bon vivants* had a firmer grasp of their wine glasses than the principles of physics:

> *Newton talk'd of lights and shades,*
> *And different colours knew, Sir:*
> *Don't let us disturb our heads—*
> *We will but study two, Sir.*
> *White and red our glasses boast,*
> *Reflection and refraction.*
> *After him we name our toast—*
> *'The centre of attraction.'*[16]

Newtonianism

Newton left a complex intellectual legacy. For one thing, he was involved in a huge range of activities: these included alchemy, biblical analysis, ancient chronology and coin production, as well as all the topics we normally associate with science, such as optics, mechanics and astronomy. Even a casual flick through Newton's two most famous books, the *Principia* and the *Opticks*, reveals that they covered a far wider range of topics than suggested by their titles. For instance, as well as the movement of the planets, the *Principia* analyses reflection, tides and how water flows out of a hole in a container. The summary of his ideas that Newton added in 1713 concentrates on the nature of God rather than physics. In the *Opticks*, Newton included discussions of volcanoes, electricity, chemical reactions, capillary action, animal structure – and, of course, God. In addition, Newton wrote several other books, now scarcely remembered, containing information about Jewish measurements, Egyptian kingdoms and cabalistic numerology.

This diversity meant that, particularly in the first half of the eighteenth century, Newton's successors pursued his ideas in different directions. Some of them, for instance, were primarily interested in his millenarian prophecies, while others were more concerned about his dating of the *Argo*'s voyage. As another example, one of the first experimenters to pursue the lines of research suggested in the *Opticks* was Stephen Hales, a country clergyman concerned to convince his readers of 'the being, power and wisdom of the divine Architect'. Opening with an account of potting sunflowers, Hales's extremely influential *Vegetable Staticks* was a careful study of plant growth that had absolutely nothing to do with optics.[17]

To complicate matters still further, Newton's books were riddled with inconsistencies, contradictions, speculations – and even plain errors. As a consequence, natural philosophers could choose to incorporate particular aspects of Newton's works into their own ideas, label themselves Newtonian, yet be promoting theories that differed substantially from one another. As Newton's fame was consolidated, it became an advantageous strategic move to pledge allegiance to the Newtonian cause, but this umbrella description sheltered different brands of natural philosophy beneath it. So-called Newtonian colleagues argued vociferously among themselves, but were united in their insistence that Newton was right. Opposition to this powerful Newtonian ideology became increasingly fruitless.

Newtonianism is a deceptively straightforward word. By 1765 it had already – according to the French *Encyclopédie* – acquired five distinct meanings. Since then, it has become even more ambiguous, as new interpretations of Newton's ideas have been generated. At its most general level, Newtonianism now signifies adopting a scientific approach not only to the physical world, but also to living beings and social systems. It was during the Enlightenment period that Newtonian natural philosophers started to advertise this emphasis on taking a rational experimental approach in order to model nature and society with mathematical laws.

Newtonianism is now pretty well restricted to scientific writings, but even among scientists it means different things. In addition to this technical splitting, the term can also refer to other aspects of Newton's thought. Because science and technology are so fundamental to the modern world, we celebrate Newton overwhelmingly for his contributions to topics such as physics, astronomy and mathematics. This was much less the case in the century after his death, when Newton was also renowned for his expertise on money, biblical interpretation and historical dating. As society changed, these other strands of his influence mostly faded away, although traces do still remain. For instance, a group called The Lord's Witnesses funds expensive advertising campaigns promoting Newton as a millenarian whizz-kid. Using his mathematical approach to biblical analysis, based on the power of 666 (see below, page 78), they can apparently predict exactly when a small clique of ten leaders will take over the world.[18]

At the beginning of the eighteenth century, it would have been hard to predict exactly which aspect of Newton's thought would ultimately prove most significant. One of his earliest biographers described a scholar who appears only distantly related to the scientific genius we commemorate today. For this anonymous author, Newton's measurements of the dimensions of Solomon's temple and the durations of royal dynasties seemed as valuable as calculating the trajectory of a bullet or the motion of the moon. Readers might well have gained the impression that although the *Principia* was undoubtedly an important book, it was possibly overshadowed by Newton's *Dissertation upon the Sacred Cubit of the Jews*.[19]

Newton's posthumous reputation was honed against those of colleagues who had shared his interests, but who were retrospectively fashioned to appear different from one another. For instance, during their lifetimes, Locke and Newton were often bracketed together as learned authors. Yet in the early nineteenth century, when the arts and sciences were separating, they were each converted into distinct cultural heroes; one French writer imagined them sitting in Oxford carving up the disciplines between them, Newton agreeing to take over the physical sciences while Locke was to lead the humanities. Similarly, Boyle had been a leading light in the early Royal Society, renowned for his chemical and pneumatic experiments. But during the eighteenth century, he gradually became distinguished for his deep religiosity, thus allowing Newton to become the hero of natural philosophy, and reinforcing the distinctions between science and religion.[20]

Ideas do not, however, travel by themselves. We can discern general trends in the past, such as the fall of Descartes, the rise of Newton, and his subsequent displacement by Einstein. Yet these massive transformations only came about because of the cumulative activities of many, many thousands of individuals who were each absorbed in their own activities. As George Eliot observed at the end of *Middlemarch*, the world is improved through small 'unhistoric acts'. During the eighteenth century, Newton's supporters promoted different facets of his thought. Focusing on his disciples reveals some of the diverse ways in which countless men (and a few scattered women) manoeuvred to improve their own positions, and simultaneously forged an impregnable Newtonian orthodoxy that came to pervade eighteenth-century life.

Three men in particular represent contrasting strands of Newton's impact and illustrate some unfamiliar aspects of how Newton's ideas have affected modern views: Willliam Whiston, George Cheyne and John Desaguliers. Each with his own fascinating life story, they are now rather obscure figures, but were renowned during the early eighteenth century as fervent champions of the Newtonian cause. Although they never met together, as members of the relatively small and intimate community of learned scholars, their trajectories overlapped as they each carved out individual paths of self-advancement. It is only in retrospect that these three individuals seem to be oriented towards three subsequent Newtonian destinies: the near-oblivion of Newton's preoccupation with ancient chronology and biblical prediction; the application of his ideas about the physical world to revolutions in biological and social thought; and the transformation of Newton's secretive quest to learn more about God's creation into giant scientific and technological enterprises.

The man who attained the greatest academic eminence was

William Whiston, who briefly held Newton's chair at Cambridge, but became notorious for pursuing those theological and historical aspects of Newton's inquiries that were squeezed out of public memory. The most widely esteemed was George Cheyne, a fashionable doctor who helped conduct Newtonian ideas into biological research and political treatises. Only one of them, John Desaguliers, an experimental lecturer and entrepreneurial engineer, dedicated himself primarily to promoting what we now value as Newton's greatest achievement, his contributions towards modern science.

Whiston, Cheyne and Desaguliers were all determined to convert Newton's arcane ideas into general knowledge, but they chose very different routes and interpreted his work in very different ways. Addison's Mr Spectator declared that 'I shall be ambitious to have it said of me, that I have brought Philosophy out of Closets and Libraries, Schools and Colleges, to dwell in Clubs and Assemblies, at Tea-Tables, and in Coffee-Houses.'[21] Like him, these three Newtonian publicists were determined to make Newtonianism a public commodity rather than a private preoccupation reserved for esoteric academics. Whiston, Cheyne and Desaguliers were unique, but they represent significant types of response to Newton's work. Their contrasting approaches illustrate how Newton's posthumous reputation was fashioned not only by scholarly university men, but also by the poets, journalists, preachers and instrument makers of Enlightenment England for whom economic survival was as strong a motivation as academic legitimacy.

William Whiston (1667–1752): millenarian crusader

It has always been tempting to poke fun at Whiston. In the Madhouse scene of Hogarth's *Rake's Progress*, an inmate of the Bedlam Hospital stares bemusedly at his sketch of Whiston's rejected scheme to measure longitude; in *The Vicar of Wakefield*, Oliver Goldsmith's comic cleric is modelled on this scholar who sacrificed his Cambridge chair for his religious convictions. In 1750, when London was shaken by two earthquakes, Whiston gave lectures denouncing the 'horrid Wickedness of the Present Age'. Quoting from the Book of Revelation he predicted further earthquakes that only the virtuous would survive. His insistence that texts in the Bible foretell the imminent arrival of Christ and the restoration of the Jews to Jerusalem made Whiston the butt of harsh Augustan wits, and invited contemporaries as well as subsequent historians to dismiss him as mad.[22]

In fact, Whiston's ideas were attuned to contemporary concerns and closely resembled those found in Newton's own writings. Whiston was no ill-educated crank: Locke's learned circle thought extremely highly of his first book, and Cambridge mathematicians appointed him the Lucasian Professor. Whiston's major eccentricity was to advertise publicly what Newton prudently kept private. This best-selling author introduced far more people to the *Principia* than had Newton himself, but he became ostracized from Georgian society by deliberately flouting religious respectability and persistently propagating aspects of Newton's theological analyses that had come to be deemed unacceptable.

The trajectory of Whiston's own life mirrors the profound transformations in attitudes towards natural philosophy that took place in the first half of the eighteenth century. The *Principia* and the *Opticks* became canonical texts, whilst Newton's researches into biblical prophecy became marginalized. But as Whiston grew older, he increasingly focused on the millenarian aspects of Newton's work. His once-glowing reputation degenerated into scornful dismissal, reflecting the move of public faith away from biblical interpretation towards philosophical experimentation as the most reliable route to knowledge about the world. Most natural philosophers were trying to gain prestige by ridiculing interpretations of abnormal events – earthquakes, comets, floods – as heralding divine vengeance. Whiston obstinately clung to what was rapidly becoming an outmoded approach, conflating apocalyptic prophesy with philosophical prediction. For Newton's successors, denigrating Whiston as a millenarian obsessive hovering on the edge of lunacy consolidated Newton's scientific credentials and at the same time minimized his unorthodox religious views. Yet despite their derision, Whiston's influence still resonates today.

Whiston's first success came in 1696 when, as an obscure provincial chaplain, he produced a highly acclaimed book; dedicated to 'Summo Viro Isaaco Newton', it was repeatedly published over the next sixty years. Its full title indicates the intimate relationship between theology and natural philosophy in the late seventeenth century: *A New Theory of the Earth, From its Original, to the Consummation of all Things. Wherein The Creation of the World in Six Days, The Universal Deluge, And the General Conflagration, As laid down in the Holy Scriptures, Are shewn to be perfectly agreeable to Reason and Philosophy*. For the previous decade, scholars and clergymen had been furiously debating whether or not natural philosophers threatened to undermine the Church's authority by contradicting biblical accounts of the earth's formation. Whiston's intervention in this heated controversy simultaneously catapulted him to fame and publicized Newton's ideas.

Inspired – like Newton – by the rash of comets that fascinated his contemporaries and were widely interpreted as messages from God, Whiston astutely reconciled God's two books, the Bible and the natural world. Mathematical calculations jostled with biblical quotations, and cometary orbits were treated with the same intellectual gravity as the Garden of Eden. According to Whiston, the words of Moses had been divinely dictated, yet provided a historical description of the world's creation that was compatible with the new mathematical and physical explanations being formulated by natural philosophers. Hypothesizing that a comet had long ago collided with the earth and triggered off Noah's flood, Whiston juxtaposed Newtonian techniques of scriptural interpretation and philosophical exposition to reinforce the reliability of both approaches.

In the same year that Whiston's *New Theory* appeared, Newton moved to London to take up his position at the Mint. As he severed his links with Cambridge, he invited Whiston to take over his Cambridge lectures for him, and approved – may even have arranged – his protégé's selection to deliver the Boyle Lectures of 1707. Funded by a bequest in Boyle's will, this annual series had been initiated in 1692 'for proving the Christian religion against notorious infidels, viz, Atheists, Deists, Pagans, Jews, and Mahometans'.[23] This frightening roll-call of enemies apparently threatening the Church of England's stability included philosophers such as Thomas Hobbes and Benedict de Spinoza. Far from being atheists by modern definitions, they were challenging traditional political and theological beliefs about the relationships between God, society and knowledge.

Many historians judge the Boyle Lectures to have played an important role in consolidating Newtonian ideology. In the troubled political atmosphere after the Glorious Revolution of 1688, most of the lecturers held what we might now describe as Protestant, progressive attitudes towards religion and politics. Deliberately calling themselves Low Church men, they drew on Newtonian ideas to promote an alliance between Christianity and rationality. These Whig adherents wanted to support the new Hanoverian monarchy in the face of continued High Church Tory support for the dethroned, yet divinely appointed, Stuart dynasty. Since their Boyle sermons were often published, they helped to consolidate a characteristically British version of natural theology, the quest to learn about God by examining nature.

Anglicans vacillated between two conflicting images of God. Newton had often portrayed a deity who intervened intermittently in the smooth running of the cosmos: thus he regarded comets as God's way of fine-tuning the celestial mechanism. This model went some way towards resolving the nagging problem of miracles, although critics were quick to point out that too many miracles would contradict the ideal of cosmic order. As Leibniz acerbically commented: 'Sir Isaac Newton, and his followers, have also a very odd opinion concerning the work of God. According to their doctrine, God Almighty wants to wind up his watch from time to time: otherwise it would cease to move. He had not, it seems, sufficient foresight to make it a perpetual motion.' British natural philosophers discarded this interventionist God to favour portrayals of a divine watchmaker whose Newtonian universe ran itself like clockwork, governed by the laws of gravity. Their paternalistic deity presided over a stable, well-regulated universe that mirrored orderly Augustan society.[24]

Newton suggested to Whiston that his Boyle Lectures could be devoted to showing how historically provable events fulfilled God's own prophecies that had been directly transcribed in the Bible. Taking this advice to heart, Whiston discovered over 300 instances, enabling him to fix the start of the apocalypse and the millennium at 1736 (later updated to 1766 when 1736 passed undramatically). Echoes of Newton's famous but ambiguous remark, 'I feign no hypotheses', resonated throughout Whiston's preaching. Both men believed that just as modern natural philosophers should reject metaphysical speculation and rely on mathematics and experiments, so too the prophecies of the Old Testament should be analysed literally rather than allegorically.

Whiston's audiences welcomed this aspect of his Newtonian biblical exegesis, but became disenchanted with his outspoken claims that historical research showed prevailing Christian beliefs in the Holy Trinity to be falsely founded. Unlike Newton, who self-protectively concealed how his chronological reinterpretations confirmed his own anti-Trinitarianism, Whiston refused to heed counsels of caution. Deviating from orthodoxy was highly risky, as a diarist's anecdote reveals: 'A lady asked the famous Lord Shaftesbury what religion he was of. He answered the religion of wise men. She asked, what was that? He answered, wise men never tell.'[25] Anti-Trinitarianism, denying God's threefold nature as Father, Son and Holy Spirit, was a legal offence punishable (in principle, at least) by imprisonment or even death. Newton astutely followed public opinion by dissociating himself from his over-ardent disciple, but, proclaiming his adherence to the unorthodox creed known as Arianism, Whiston defiantly continued to question prevailing beliefs in Christ's holy status. As this idealistic scholar naively ventured into the dangerous terrain of Church politics, he became caught up in the wake of a national scandal. Banished from Cambridge for heresy, he fled to London.

Distinguished metropolitans like Addison welcomed this fervent Newtonian. They felt that if his 'Itch to be venting his Notions about Baptism & the Arian Doctrine' could be kept under control, his track record at Cambridge suggested that he would be a profitable recruit to teach the new courses in natural philosophy. Whiston was a successful and entertaining lecturer as well as a prolific author, one of the earliest and most significant popularizers of Newton's mathematics and natural philosophy. One particularly enthusiastic student was Alexander Pope, whose poems ensured that Newton's fame reached audiences far beyond those frequenting London halls and coffee houses.[26]

Produced in long and cheap print runs, books like *Sir Isaac Newton's Mathematick Philosophy More Easily Demonstrated* brought Whiston a much-needed income, and also converted Newton's complicated theoretical speculations into simple unshakeable certainty. Ignoring Newton's grumbles, Whiston resuscitated some of his earlier papers on algebra to publish them as *Arithmetica Universalis*, a deliberate riposte to Descartes's book on geometry. Translated into English, this became the century's most widely read book on Newton's mathematics; forty years later, Newton falsely boasted that he had chosen the title himself.

But particularly after Newton's death, Whiston publicized aspects of Newton's thought that have become less familiar. Theological issues were central to Newton's life and thought, even though only privileged friends were aware how far his beliefs deviated from orthodoxy. Over the past three centuries, we have come to celebrate Newton as a genius of science, and few people are aware of his writings on ancient chronology, scriptural interpretation and prophecy, which survive in voluminous manuscripts and posthumously published books. These were not simply casual interests to fill a Sunday afternoon, but fundamentally affected what we call his scientific work.

How, Newton repeatedly asked himself, can a spiritual God

interact with the material world? For help, he immersed himself in ancient alchemical and scriptural texts, so that his interest in unorthodox spiritual beliefs moulded his constantly shifting speculations about aether and the operation of gravity. His dedication to alchemical experimentation was no vain quest to transmute lead into gold, but a committed search for an aetherial animating spirit, evidence of God's presence in the universe. As Newton delved into the writings of the Greeks, he became convinced that his natural philosophy was a retrieval of lost ancient knowledge. He argued, for instance, that the Babylonians had – like him – placed the sun at the centre of a universe composed mainly of empty space. As Colson's frontispiece advertises, Newton preferred to use traditional geometrical arguments (*Figure 3.1*). He turned his back on the new algebraic methods being developed on the Continent, thus adversely affecting British mathematics during the eighteenth century.[27]

Newton, like many of his contemporaries, held ideas that now often seem outlandish, such as his conviction that Nebuchadnezzar's dream of a monstrous image made from four metals foretold the four successive monarchies of Babylon, Persia, Greece and Rome. He scoured Ezekiel to determine the dimensions of Solomon's temple, relied on his detailed knowledge of the Apocalypse and breeding patterns of locusts to redate historical events of the Roman Empire, and confirmed Christ's future eternal reign by translating words into their numerical coded equivalents: most famously, 666 was the number of the Beast.[28] Anxious to ensure that he did not meet Whiston's fate, Newton went to considerable lengths to conceal his heretical theological views, but information did leak out. Unauthorized copies circulated of a manuscript on ancient chronology that Princess Caroline had forced him to write for her. She often boasted to her friends that it was one of her most treasured possessions, and the version of Kneller's portrait that she owned shows Newton with a Greek copy of the Book of Daniel (*Figure 2.1*).

Clergymen and natural philosophers joined forces to reinforce their authority over the uneducated, and to displace older authority figures such as astrologers. God, they insisted, ruled through natural laws, and it was just superstitious rubbish to interpret unusual astronomical phenomena as religious portents. But, convinced that divine significance could be read into unusual events like comets and eclipses, Whiston continued to promote Newton's own commitment to treating philosophical prediction and biblical prophecy as two sides of the same coin. Delivering lectures that mixed astronomy with the restoration of the Jews to Jerusalem, Whiston toured the country displaying scale models of Moses' tabernacle and Ezekiel's temple. He earned a small fortune from lectures and booklets that used detailed mathematical diagrams to predict and explain the total solar eclipse of 1715, but he also regarded unanticipated comets, meteors and unprecedented appearances of the northern lights over London as explicit warnings from God. He managed to secure private as well as public funding for his research into magnetism, but devoted pages of complicated calculations laced with biblical quotations to showing how his measurements of the earth's changing magnetic patterns confirmed scriptural chronology.[29]

Probably irked by Newton's personal repudiation, Whiston attacked Newton's biblical interpretations. This opposition fanned international interest, and eighteenth-century commentators – including Voltaire and Edward Gibbon – took Newton's ideas on chronology very seriously. One critic marvelled that the 'vastness of his Genius extended to every part of Literature . . . he has entred on methods for fixing of Epocha's never before thought of by others'. Correspondence flowed, and heated arguments took place in coffee houses, private dining rooms and university studies. Thousands of pages were published on Newton's attempts to redate the expedition of the Argonauts by combining astronomical calculations with analyses of fragmentary Greek poetry. Newton was cited as a biblical and chronological expert by men who are now misleadingly celebrated only as founding fathers of scientific disciplines, such as Joseph Priestley for chemistry and David Hartley for psychology. This one-sided commemoration commits to obscurity many volumes of writings contemporary readers regarded as important.[30]

For scientific polemicists, Newton's absorption in such niceties as the alchemical green lion, the seven vials of the angels and cabalistic numerology became embarrassing. These preoccupations had dominated much of Newton's life, but they detracted from the aura of rationality appropriate for a modern intellectual hero, and during the nineteenth century they became sanitized as 'relaxations in mature life from hard thinking and investigation'. Hagiographers conveniently glossed over the details of Newton's anti-Trinitarian beliefs, and instead praised him as a deeply religious man. When David Brewster, Newton's major nineteenth-century biographer, examined Newton's correspondence, he reportedly decided that 'of theological papers, only such will be published as are sufficient to *prove that Newton believed strictly* in the Trinity'.[31]

Whiston had singled out for approval those aspects of Newton's writings that confirmed his own Unitarian insistence in God's unique holiness, a belief regarded as heretical because it denied the conventional Trinitarian doctrine of God's threefold nature as Father, Son and Holy Spirit. Unitarianism attracted a growing number of adherents. Early members included Priestley, the Birmingham chemist, and also Thomas Jefferson, who greatly admired Newton. He sent to London for a special portrait to hang in his office, used Newtonian mathematics to design a new plough, and rewrote the New Testament by excluding references to Christ's divinity.[32]

As British Unitarians became more firmly established in the nineteenth century, they 'gloried in his [Newton's] authority, & appropriated him to themselves'. Even those opposed to Unitarianism regarded Newton as an important target: one book was triumphantly subtitled *Sir Isaac Newton and the Socinians Foiled*. Far from dying out, millenarian prophesy flourished amid the uncertainties of the American and French Revolutions, and modern fundamentalist movements now venerate Newton as a founding father. *March of the Reformers*, a picture produced by the Seventh Day Adventists, depicts Newton passing on the torch of knowledge from the past to the present. But unlike galleries of scientific innovators where he follows Kepler and Galileo, here Newton heads a line of heroic scriptural interpreters such as John Knox and Martin Luther, leading back to Daniel himself.[33]

Newton's *Observations upon the Prophecies of Daniel, and the Apocalypse of St John* has been repeatedly republished. In 1922, Sir William Whitla, one-time head of the British Medical Association and enthusiastic supporter of the Salvation Army, produced his new edition to counter scriptural scholars who regarded the Bible as a collection of books written by men rather than God. Guided by Newton's methodology, he explained how Daniel had foretold the then recent occupation of Jerusalem by British troops. The book appeared again as recently as 1991. In his introduction, Arthur Robinson, a modern fundamentalist physicist associated with the environmental movement in Oregon, explained that 'Like Isaac Newton, I do not know of any verified scientific facts that are inconsistent with the literal truth of every aspect of the Bible.' Newton has proved particularly valuable for groups whose calculated end of the world has repeatedly failed to materialize, since he maintained that it was impossible to pin down the dates of prophecies precisely. In 1999, the Web still carried a series of calculations reporting that if 'Newton was correct in his interpretation of Daniel . . . then 1996 or 1997 will be the year of the return of Jesus Christ'![34]

Jerusalem is now the home of many of Newton's manuscripts. This rather surprising location was chosen by the scriptural expert A. S. Yahuda, who shared Newton's insistence on the Bible's historical accuracy. Yahuda's bequest of Newton's theological manuscripts has enabled scholars to learn far more about Newton's beliefs and analytical techniques. In 1990, after an American historian pointed out that some of Newton's interpretations seemed to predict key dates in the formation of modern Israel as a Jewish state, a fundamentalist in Jerusalem telephoned him for further information. Whiston would have approved.

George Cheyne (1671–1743): medical reformer

William Hogarth delighted in visual and verbal *double entendres*. One good example is his theatrical scene of Conduitt's drawing room (*Figure 2.7*), which is saturated with Newtonian imagery. Another is his hallucinatory image of *The Weighing House Inn*, in which Hogarth punningly plays on gravity, with its dual connotations of physical weight and mental seriousness (*Figure 3.2*). Hogarth designed this ranked array of levitating intellects for the frontispiece of a humorous pamphlet by John Clubbe, a country rector who, enjoying a local reputation for satire, had elaborated on some lines from Pope's *Dunciad*:

> *Hear you! whose* graver *heads in equal scales*
> *I weigh, to see whose* heaviness *prevails;*
> *Attend the trial I propose to make.*

Clubbe had suggested that a magnetic weighing machine be set up in every town to 'prevent great impositions on the publick; for, if the solid contents of every man's head can thus be come at, every one will know how far he may trust to the understanding of his neighbour'. Choosing the forecourt of an inn, a favourite setting for revealing social transgressions, Hogarth has doubly inverted daily order by dressing the foolish light-headed men in far better clothes than the working men of good sense, who are standing on their heads.[35]

This picture is sometimes cited to demonstrate how thoroughly Newtonian concepts pervaded eighteenth-century thought. As with many Georgian witticisms, Clubbe's and Hogarth's sense of humour now seems rather alien. The failure to appreciate jokes generally betrays profound cultural clashes, and the strangeness of this gravitational caricature highlights some of the ways in which people thought and felt differently from us. Although we have come to demarcate the arts from the sciences, a quarter of a millennium ago the intellectual and emotional world was far less segmented. We give priority to logical, deductive chains of reasoning, but many philosophers argued by analogy from one sphere of experience to another. Language resonated with multiple metaphorical meanings, as single words transported implications from one context to another. The boundaries separating scientific, ethical, economic and religious issues were highly permeable, permitting a fluid interchange that encouraged analogical reasoning, a sort of rational equivalent to the Hogarthian-style visual and verbal puns so relished by learned Augustans.

The physician George Cheyne, who was greatly renowned for his tracts on diet and medicine, shared and also literally embodied Hogarth's fascination with Newtonian gravity and analogical modes of argument. 'The learned Doctor Cheyne' (as he appeared in Henry Fielding's novel *Tom Jones*) exemplifies how Newtonian ideas could diagnose bodily, spiritual and social ailments. As his lifestyle vacillated between gregarious gluttony and secluded abstemiousness, his weight swung repeatedly between a 'Lank, Fleet and Nimble' 130 lb (59 kg) and an 'excessively *fat, short-breath'd, Lethargic* and *Lifeless*' 448 lb (204 kg).[36]

These bodily fluctuations corresponded both to his psychological transformations and to his changing attitudes towards Newtonian attraction. Like many of his contemporaries, Cheyne believed that God harmoniously governed His entire creation with simple laws, so that powerful analogies shimmered between the ways in which particles were bonded into matter, individuals cohered in social groups, or souls attained reunion with God. Cheyne's personal movements between conflicting models of gravity reflected the ambiguity of Newton's own statements as well as the waxing and waning of competing interpretations during the century.

Newton had bequeathed sketchy and contradictory conjectures about how gravity might operate. The most apparently straightforward of his views, and the one reinforced by the second edition of the *Principia* that appeared in 1713, held that gravity acts at a distance without any intervening medium. But that hypothesis smacked too much of ancient occult powers, the quasi-magical forces that natural philosophers claimed to have dispelled through their rational approach towards nature. Moreover, many religious believers objected that making particles inherently forceful removed the distinction between inert matter and God's spiritual power. They preferred to envisage some sort of divine spirit driving a cosmic machine, in which inert particles moved only because they were directly pushed.

As Newton dodged his critics and developed his alchemical ideas, he also investigated the notion of a subtle spiritual aether. To explain phenomena such as optical refraction and electricity, he tentatively suggested that particles might have both repulsive and attractive properties, their influence varying with size. Perhaps, he conjectured, an aetherial fluid made up from tiny repellent particles pervades the whole of space, a medium able to transmit gravity and magnetism, yet one so rare that it scarcely affects the motion of the planets.

Like Newton, Cheyne adopted different models. Initially, he made divine attraction analogous to the central tug of gravitation. But later, he compared spiritual power with the short-range forces that bind particles together into matter, describing people as reflections of God, 'diminutive analogical *Particles*' whose pull towards each other paralleled the attraction drawing them towards Him.[37] By the end of Cheyne's life, as natural philosophers constantly reinterpreted Newton's ideas, many of them had come to share his focus on aethers made up of small particles as the medium of attraction.

Like Whiston, Cheyne was a mathematical millenarian and an early Newtonian who was later rejected by Newton himself. But there the similarities end. Far from being ridiculed, his enormously successful books on dietary regimes and the ills of English society made him somewhat of a cult figure. In contrast with the mainly Low Church Whig leanings of the Boyle lecturers, Cheyne belonged to a small group of predominantly Tory Scottish Episcopalians. They clustered around two distinguished Edinburgh academics, David Gregory and Archibald Pitcairne, ambitious Newtonian proselytizers whose careers were shadowed by their Jacobite leanings. With Newton's help, Gregory managed to flee to a safe professorship at Oxford, which had been staunchly Royalist during the Civil War and remained a Tory stronghold well into the nineteenth century.[38] Cheyne himself emigrated to London in 1701, but continued to divide his energies between the mysticism, mathematics and medicine he had absorbed in Scotland, a heady mixture whose inherent compatibilities subsequently dominated his conflicted life.

Although Cheyne never did write his proposed *Mathematical Principles of Theoretical Medicine* (yet another *Principia*!), he was one of the earliest popularizers of Newton's theories of attraction. His first book, on fevers, defended what Pitcairne labelled 'iatromathematics', an influential school of thought that flourished until the 1730s. Satirically dubbed 'the Art of Curing Diseases by the Mathematicks', this Newtonian approach to medicine and physiology searched for the facts of life in the short-range attractive forces between particles of matter. It was one version of the new mechanical models of the body that natural philosophers had been introducing as they adopted a quantitative and experimental approach to physiology.

Traditional physicians retained a more holistic view. Drawing no firm distinction between bodily and spiritual ailments, they believed that people's behaviour and health were governed by the balance of four humours. In contrast, the iatromathematicians conceived the body as a hydraulic system filled with moving fluids whose behaviour could be explained by using Newtonian mathematics. As Cheyne explained, 'the Human Body is a Machin of an infinite Number and Variety of different Channels and Pipes, filled with various and different Liquors and Fluids, perpetually running, glideing, or creeping forward, or returning backward, in a constant *Circle*'.[39]

Cheyne continued to publish Newtonian books on mathematics and medicine after he came to London, but things soon started going disastrously wrong. He effectively ostracized himself from the Royal Society by antagonizing Newton and his supporters, who accused him of plagiarism and mathematical incompetence.

Renouncing his former abstemious industriousness, over the next few years Cheyne ate and drank himself to breaking point. As he later chastised himself, because he frequented coffee houses and taverns to drum up medical business he over-indulged in 'luxury, gluttony, and upper-class vice without exercise', constantly 'taking snuff out of a ponderous gold box'. At the age of thirty-five, weighing 440 lb (200 kg), scarcely able to let go of 'the Posts of my Bed, for fear of tumbling out' and convinced that he was about to die, he withdrew to the country. There he resolved to purify his body with purges, emetics and a strict vegetarian regime, and to cleanse his soul through spiritual reflection and religious reading.[40]

Probably under Whiston's guidance, as Cheyne ruminated in his rural retreat, he came to see his own restoration to health as a divine sign that he should help to establish the New Jerusalem in England. No longer drawn by London society and luxurious living, he experienced a cosmic force that pulled him towards God just as gravity bound the planets to the sun. God, he explained to his readers, had imbued living creatures with 'a *central* Tendency towards Himself, an Essential Principle of Reunion with Himself, *Analogous* to this principle now mention'd [gravity] in the Great Bodies of the Universe'. Like comets disturbing the regular motion of the solar system, so too 'earthly and sensual *Attractions* . . . destroy the beautiful progress of *spiritual Beings*, towards the Centre and End of their being'.[41] While Cheyne argued by analogy to translate Newton's laws of gravitational attraction from the physical universe to the moral one, his own weight diminished as he rid his body and his soul of impurities.

In 1709, Cheyne felt ready to face the world and resume his medical practice in Bath. But for the next fourteen years, although he was busily disseminating his new concepts of attraction, continuing his religious studies and publishing books on healthy living, he struggled incessantly against consuming temptations. Periodic stints of vegetables and medicinal waters proved insufficient to counter his alcoholic calorie intake, and his weight escalated until a servant had to follow him with a stool so that he could recuperate every few yards. Medical treatment often lags behind physiological theories, and although Cheyne had endorsed Newtonian iatromathematics, he prescribed older therapeutic techniques of dietary regulation. With the help of a lettuce-and-wine regime he eventually managed to move into a more settled final phase of his life. Still pursuing his religious introspection, he shrank to a third of his former mass, probably encouraged by the huge success of his controversial books about personal well-being and the health of the English nation. Cheyne became a fashionable physician who dispensed medical and moral prescriptions to aristocratic women and some of the country's most famous men – writers like David Hume, Samuel Richardson and Alexander Pope.

Cheyne was one of the first of Newton's successors to explore aether models, which became increasingly prevalent from around 1740. Interpretations varied enormously, largely because as the mediators between matter, motion and spirit, aetherial fluids carried huge theological implications. Relying on arguments that ranged from the ineffably vague to the extraordinarily convoluted, natural philosophers described weightless invisible fluids of subtle particles seeping through the pores of solids, forcing gases to expand, and cushioning the sun in a great repellent cloud whose graduating density maintained the planets in their appropriate orbits. Often authenticated by the adjective 'Newtonian', aethers proliferated and diversified as authors with very different religious commitments summoned them up to explain mysterious phenomena like electrical charge, magnetic repulsion, or human memory.[42]

Cheyne's interest in investigating how Newton's ideas could cast light on human behaviour was widely shared. By far the most influential of his successors was David Hartley, an ardent proponent of Newtonian aethers. Although he wrote extensively on the millenarian beliefs he shared with Newton and Whiston, Hartley became widely celebrated for his psychological innovations. Like Cheyne, he was an intensely religious physician who argued analogically and sought to embrace physical and moral realms within a single philosophy. Set out in numbered propositions like the *Principia*, Hartley's famous *Observations on Man* (1749) covers the physical causes of itching, yawning and sexual arousal as well as life's moral pains and pleasures. According to Priestley, Hartley threw 'more useful light upon the theory of the mind than Newton did upon the theory of the natural world', while Coleridge expressed his enthusiasm by naming his son Hartley.[43]

Just as Cheyne had converted Newton's mathematics into hydraulic physiology, Hartley elaborated one of his speculations in the *Opticks* into an aetherial model of the nervous system. Light falling on the eye might, Newton had suggested, set up vibrations that travel along the nerves to the brain. Hartley envisaged human brains and nerves as filled with a subtle fluid that transmits signals from one part of the body to another. An external stimulus – such as pain, smell or sexual desire – causes characteristic vibrations in this internal aether, which give rise to specific responses in the brain. By combining this physical model derived from Newton with contemporary arguments that ideas become associated in people's minds when they occur close together, Hartley provided new ways to think about free will, ethics and biology.

Hartley's associationism, as his ideas came to be called, had a huge impact on theories about physiology, psychology and evolution, and also on moral philosophy. In economics, Adam Smith was arguing that individual profit could be compatible with public wealth. Analogically, Hartley described psychological and physiological checking mechanisms that enabled people driven by self-interest to participate in mutually beneficial social enterprises. In other words, he made personal satisfaction compatible with the greatest public happiness – the ultimate goal of the English utilitarians who followed him.

Observations on Man was enormously influential in France, where Newtonian principles came to govern the biological and moral realms as well as the physical universe. Focusing on the social implications of Hartley's work, leading philosophers such as Étienne Bonnet de Condillac and the Marquis de Condorcet constantly referred to Newton's innovations as they tried to deduce general laws describing human behaviour and morality. In parallel endeavours, men of science like Georges Buffon (who gazed for inspiration at the portrait of Newton above his desk) continued Cheyne's and Hartley's quest to regulate the living world with Newtonian laws of physics. 'Let us render homage to Newton,' urged one of France's leading medical researchers in 1801, the 'first to discover the secret of the Creator, viz. a simplicity of causes reconciled with a multiplicity of effects.'[44]

John Theophilus Desaguliers (1683–1744): entrepreneurial lecturer

In Hogarth's picture of Conduitt's Newtonian drawing room, a short, stocky clergyman with his back to the audience peers into the play's text to prompt the childish actors (*Figure 2.7*). John Theophilus Desaguliers appears here as Newton's protégé, a man of reason maintaining Augustan order amidst frivolous revelry. But Hogarth was not always so gentle. His audiences inside and outside this conversation piece knew that in other performances, this sombre off-stage assistant paraded in ceremonial splendour as one of the country's leading Freemasons. In savage caricatures, Hogarth portrayed Desaguliers as a myopic preacher boring a lascivious clergyman and a dozing congregation, and as an old woman baring her buttocks in a Masonic procession.[45]

Despite gaining international eminence both as Grand Master of the newly founded Grand Lodge of England and as Fellow of the Royal Society, Desaguliers had a hard life. A refugee from religious persecution (as a toddler he had been smuggled to safety in a barrel), he struggled incessantly to support his family. Scientific inventors had not yet carved out what would prove to be one of their most valuable inventions: the scientific career.

Frustrated at feeling himself the Royal Society's hired servant, Desaguliers constantly supplemented his meagre salary with a variety of speculative ventures in publishing, engineering and lecturing. He converted his house into a boarding school, but was forced to leave when it was demolished to make way for Westminster Bridge, even though he had been one of the consultants employed on this massive project. Like many corpulent Georgian gentlemen, Desaguliers suffered from chronic gout (most probably turning for advice to Cheyne's best-selling self-help manual), and he died in poverty at his lodgings in Covent Garden.

Although an easy target for Hogarth's wit, Desaguliers was an enterprising and well-educated man, equally at home in French, English or Latin. As he manoeuvred among churches and coffee houses, lodges and lecture theatres, this philosophical engineer developed new machines but also engineered new sources of income. His role in promoting Newtonian ideas was singularly important, yet he also typifies the new spirit of commercial enterprise that was developing in the first half of the eighteenth century. Countless contemporaries were learning to rely on their intelligence rather than their inheritance, and their individual entrepreneurial activities contributed to establishing science as a worthwhile practical enterprise. Gowin Knight, for instance, was a doctor from an impoverished provincial family who cornered the naval market in compasses, wrote a Newtonian book about attraction and repulsion, and became the first director of the British Museum; James Ferguson acquired almost mythical status as a Scottish autodidactic astronomer who built up a flourishing instrument trade. Simultaneously benefiting from and contributing to Britain's booming economy, such self-made men collectively consolidated public approval of Newtonian natural philosophy.[46]

Desaguliers was only seven when he arrived in London, one of the Huguenots forced to emigrate in 1685 after Louis XIV revoked French anti-discriminatory legislation. Unenthusiastic about following his father into the Church, as an Oxford undergraduate Desaguliers initiated his Newtonian career by translating French engineering books and attending the extra-curricular experimental demonstrations being run by John Keill. This Scottish High Church mathematician, sneered at as 'Newton's toady', belonged to the same circle as Cheyne and spearheaded Newton's challenge to the English educational system; however, despite Keill's innovations, Oxford remained committed to Aristotelian philosophy long after Cambridge was teaching Newtonian physics. Keill probably smoothed Desaguliers's path as he insinuated himself into the competitive, commercial world of London lecturing, where he too became renowned as an ardent Newtonian propagandist.

Newton made a wise decision when he agreed to sponsor Desaguliers and secure his rapid promotion to become the Royal Society's experimental demonstrator in 1716. An energetic writer, lecturer and international traveller, Desaguliers became the most influential early propagator of Newtonian ideology amongst the international scholarly community as well as a broader public. Constantly negotiating between different social networks, he brought Newton to huge audiences, both through his own work and also indirectly through his former students, several of whom themselves became influential educators.

Benefiting from his practical skills, Desaguliers excelled at inventing apparatus specifically devised to demonstrate Newtonian principles, and his experimental demonstrations proved extraordinarily convincing. Whereas Newton had prudently cloaked his more controversial suggestions under the guise of speculation, Desaguliers confidently claimed that the *Opticks* contained a 'vast Fund of Philosophy; which (tho' he has modestly delivered under the Name of *Queries*, as if they were only Conjectures) daily Experiments and Observations confirm'.[47] Thus as Desaguliers defended, interpreted and developed Newton's suggestions, he enabled Newton's own vacillations to be conveniently overlooked and made his philosophy more palatable by imbuing it with the simple certainty of truth.

The roll-call of his children's aristocratic godparents testifies to Desaguliers's expertise in the art of cultivating patronage, so essential for survival in the cut-thrust metropolitan world. He pursued several lines of contact simultaneously. Assiduously securing Newton's favours at the Royal Society, he also took advantage of his theological training to obtain Church posts from aristocratic and royal backers, while Masonic ceremonies offered profitable encounters with wealthy gentlemen eager to invest in engineering projects. For twenty-five years, his most valuable patron was James Brydges (later the Duke of Chandos), owner of a large estate in Middlesex. In a symbiotic relationship that soured sadly as Desaguliers neglected his religious duties, Chandos offered salaried clerical sinecures in exchange for improvements to his property and technical advice on his risky financial ventures.

Desaguliers's allegorical poem *The Newtonian System* – which could aptly be retitled *The Patronage System* – illustrates his oily diplomatic expertise. He composed it to celebrate George II's coronation in 1727, the year of Newton's death, but dedicated it with blatant flattery to the new Queen Caroline, hoping that she would continue to employ him as her experimental entertainer. Deftly interweaving praise for Newton's 'tow'ring Genius', English freedom and royal splendour, Desaguliers patriotically twinned Newtonian astronomical certainty with Hanoverian stability, tactfully not mentioning that the new King had been over thirty when he arrived in his adopted country, and was not even fluent in his subjects' language. Unlike the French regime that ruled by fear and was subject to the turbulent 'Whims of the Cartesian scheme', the British King – enthused Desaguliers – resembled the sun, tranquilly extending the power of love over a nation of free citizens:

> *That* Sol *self-pois'd in* Aether *does reside,*
> *And thence exerts his Virtue far and wide;*
> *Like Ministers attending e'ery Glance,*
> *Six Worlds sweep round his Throne in Mystick Dance . . .*
> *ATTRACTION now in all the Realm is seen*
> *To bless the Reign of GEORGE and CAROLINE.*[48]

Embellished with didactic notes, astronomical diagrams and advertisements for his lecture courses, Desaguliers's *Newtonian System* was just one of the numerous poems published throughout the century that simultaneously reinforced Newton's reputation as an English genius, consolidated national familiarity with the tenets of Newtonian natural philosophy, and promoted their authors' diverse interests. In Desaguliers's case, the diplomatic strategy worked. He was soon appointed Chaplain to Frederick, the new Prince of Wales, regularly visiting Kew to give him private lectures in natural philosophy, and, a decade later, presided over the ceremony initiating him as a Freemason.

Desaguliers's salary at the Royal Society was tied to his productivity, which goes some way towards explaining his prolific publication record of over fifty papers in the *Philosophical Transactions*. These covered a huge range of topics. While Newton was still alive, Desaguliers sensibly dedicated most of his energies to defending contested items of Newtonian doctrine. For instance, he acted as Newton's Parisian ambassador, refuting French criticisms about Newton's optical experiments and countering claims that Newton had been wrong to describe the earth as being flattened at the poles (French expeditions to settle this prolonged debate eventually proved Newton to be right, a key conclusion encouraging the continental spread of Newtonian ideas).

But after Newton died, Desaguliers was freer to pursue his own inclinations, and in his later articles took Newtonian ideology in new directions. He was particularly interested in exploring the suggestions Newton had put forward in the *Opticks* about the attractive and repulsive forces between particles. Initially he focused on investigating how gases expand and contract, a topic in which he had gained huge practical expertise with his machines designed to rid city air of its noxious 'fuliginous vapours, arising from innumerable coal fires, and stenches from filthy lay-stalls and sewers . . . in the country a serene dry constitution of the air is more exhilarating than a moist thick air'.[49]

He also leaped on to the most fashionable philosophical bandwagon – electricity. Machines that generated static electrical charges had originated as by-products of Newton's research into glass, and literally shocked Enlightenment men and women into a surge of enthusiasm for spectacular natural philosophy. At dinner parties, gentlemen set ladies' drinks on fire with their electrified swords or thrilled to exciting kisses from the lips of a charged-up Venus. In lecture theatres, performing philosophers attracted feathers from the floor with the hands of electrified orphans suspended by ropes from the ceiling. Throughout Europe, learned scholars were as fascinated as everyone else by this exciting new phenomenon. In his prize-winning book, published in French as well as English, Desaguliers advertised the value of a Newtonian interpretation.

Through his theoretical discussions of gases and electricity, Desaguliers contributed to the debates among Newton's followers that resulted in the general adoption of aether theories after about 1740. However, his practical expertise also promoted Newtonian ideas. Desaguliers dispensed advice on improving Edinburgh's water-supply problems, ventilating the stuffy House of Commons, importing chemicals from Africa, exploiting Scottish pine forests – indeed, there were few problems of eighteenth-century life that he failed to tackle. He endeared himself to his wealthy private patrons by curing their smoky chimneys, heating their draughty houses and pumping water out of their mines. At the same time, he impressed the Fellows at the Royal Society and the students at his lecture courses by converting the working models of these engineering schemes into Newtonian experiments.

Profit and natural philosophy went hand in hand. In *Gulliver's Travels*, Swift satirized this opportunistic entanglement by giving his Academy of Lagado, that acid critique of the Royal Society, the physical dimensions of the Royal Exchange. Like other entrepreneurial inventors, Desaguliers became involved in bitter and costly patent disputes because he was as interested in marketing his barometers, hydrometers and other philosophical instruments as in their potential to yield new information about the world. Drawing on his research experience, Desaguliers advised the City investors he met at Masonic meetings about their speculative projects in soap manufacturing, mining and brewing. Ambitious sponsors employed Desaguliers to install ornamental fountains in their country estates, or organize impressive firework performances as ostentatious evidence of their wealth and status. Inevitably, things didn't always run smoothly: on one occasion, the rockets that should have shot up out of the Thames to explode dramatically in the sky went astray and blew a hole in the bottom of a nearby barge carrying spectators, who had to be ignominiously rowed ashore.

As Desaguliers travelled round Britain, France and Holland on Masonic missions, lecturing tours and engineering site trips, his wife was also experiencing some of the harsher aspects of Georgian life. Like so many philosophical partners, little evidence survives of Joanna Desaguliers's existence, but five of her seven children died in infancy, and she presumably bore the responsibility of caring for the paying boarders who attended her husband's courses (and who, one wonders, took over when he was laid up every winter with his chronic gout?). Although we can now only speculate about wifely contributions to the consolidation of Newtonian ideologies, these invisible assistants undoubtedly performed countless mundane chores.

In addition to the sick babies and students, Desaguliers's cramped house in Westminster must have been overflowing with his apparatus – instruments in various stages of completion, piles of unsold books, and bulky demonstration equipment that included an eight-foot wide centrifugal bellows and a working steam-engine. Desaguliers was particularly proud of his large mechanical planetarium, which he claimed displayed the movement of the planets round the sun far more accurately than ornate orreries like the one portrayed in Wright of Derby's picture (*Figure 1.4*).

The lectures, books and apparatus that were generated in this squashed household did far more to consolidate a Europe-wide Newtonian ideology than the learned disquisitions emanating from the Royal Society. In 1734, as Desaguliers embarked on his 121st course of lectures, he boasted that of the dozen or so other experimental lecturers in the world, eight had been those whom he had taught. One of them, Stephen Demainbray, later became the custodian of George III's astronomical observatory at Kew. His impressive demonstration equipment, much of it similar to Desaguliers's own, has been preserved and can now be seen in London's Science Museum.[50]

From a Newtonian perspective, the most influential of Desaguliers's former students was Willem 'sGravesande, who became a professor at Leiden, one of Europe's leading universities. Uniting optics, astronomy and mechanics in a single book, 'sGravesande made Newton's mathematics accessible by making it experimentally visible. Taking advantage of traditional Dutch woodworking skills, he devised demonstration equipment to display mechanical and optical principles with clever devices – a cone that appears to run uphill, a shadowy monster's head projected on to a wall, a miniature tower that tips over. Restyled in metal and plastic, some of these are now marketed as executive toys with appealing names, such as Newton's cradle. Ironically, thanks to Desaguliers's English translation, this modified Dutch Newtonianism was exported back into England where it dominated English teaching for many years.

Desaguliers's courses on mechanics owed more to his involvement in entrepreneurial engineering projects than to his careful perusal of the *Principia*. Financial arguments pack more persuasive punch than mathematical ones, and for many people, the effectiveness of the numerous pumps, ventilators and irrigation pipes that Desaguliers installed provided the most convincing evidence that Newtonian philosophy was worth studying. Gullible investors did of course lose their money, and Desaguliers carefully spelled out the intrinsic performance limitations of his machines as 'a very necessary Caution; for there are several Persons who have Money, that are ready to supply boasting Engineers with it, in hopes of great Returns and especially if the Project has the Sanction of an Act of Parliament to support it – and then the Bubble becomes compleat, and ends in Ruin'.[51] By helping to produce a nation well versed in Newtonian ideas and dependent on Newtonian machines, Desaguliers encouraged the mutually profitable alliance of commerce and natural philosophy that underpinned England's development into the world's first industrial economy.

4

ENEMIES

The Horse of Intellect is leaping from the cliffs of Memory and Reasoning; it is a barren Rock: it is also called the Barren Waste of Locke and Newton.

William Blake, *A Descriptive Catalogue of Pictures* (1809)

Newton was not above twisting the evidence to make the figures match his theories. The *Principia* persuasively argued that his cosmology was based on precise calculation and measurement – hence the word 'mathematical' in its title. But when the book first appeared, Newton realized that numerical discrepancies in the fine details of his system threatened its acceptance. As he battled to refute his continental critics – especially the detested Leibniz – Newton revised his calculations of the velocity of sound. Choosing a quiet arcade in Trinity College, he repeated his own measurements on echoes and manipulated the conveniently vague 'crassitude' of air particles until his recorded observations corresponded unbelievably closely to his theoretical predictions. In addition, discreetly massaging records of lunar distances and tidal heights, he instructed the overworked editor of the *Principia*'s second edition how far to 'mend the numbers'.[1]

Despite this ambivalent attitude towards philosophical fiddling, Newton became notorious for his determination to stamp out financial forgery. At the Mint, 'that old Dogg the Warden' relentlessly pursued coin counterfeiters to the gallows by collecting testimony from hundreds of witnesses, hiring spies to penetrate underworld networks, and rewarding informers who saved their own lives by naming their friends.

Life was not so very different in intellectual circles, as antagonists accused each other of doctoring observations, plagiarizing books and stealing new inventions. Loyal acolytes reported that Newton 'conversed chearfully with his friends assumed nothing & put himself upon a level with all mankind'. In contrast, his critics portrayed a scheming, secretive opportunist who trampled on opponents and bribed supporters to help fashion his reputation as the world's greatest natural philosopher. Establishing himself as the leading searcher after truth did not necessarily entail telling the truth.[2]

According to one of Newton's hagiographers, he 'had a particular aversion to disputes, and was with difficulty induced to enter into any controversy'.[3] Such posthumous whitewashing conveniently ignored accounts of Newton's numerous vituperative conflicts. One of his bitterest enemies was Robert Hooke, a gifted experimenter who constantly struggled to protect his inventions and his ideas against intellectual piracy. That there is no surviving portrait of Hooke corroborates cruel barbs about his physical appearance, but also reflects his relatively low social status. Unlike the wealthy gentlemen with whom he worked, Hooke was an employee at the Royal Society, the Curator of Experiments appointed to contrive experimental trials and entertaining demonstrations.

Hooke supervised a host of assistants who were essential for the Society's activities, but were acknowledged only fleetingly in written records. Just as these behind-the-scenes helpers have become virtually invisible, it now seems clear that while Hooke's contemporaries paid his salary, they effectively diminished his posthumous reputation by failing to credit his vital contributions to the new techniques of experimental philosophy. Allegedly a reclusive, workaholic insomniac, Hooke displayed many characteristics that might have earned him the label of genius had he occupied a less menial role and not been so obviously concerned to profit from his inventions. For thirty years, until his death in 1703, Hooke repeatedly accused Newton of appropriating theories that he had himself originated.[4]

Newton professed to welcome criticisms, but he never forgave Hooke for pointing out a fundamental error in one of his arguments, an injury he nursed for decades. At first they exchanged civil expressions of interest in each other's work, and it was in one of Newton's politer self-defensive letters to Hooke that he famously declared, 'If I have seen further it is by standing on ye shoulders of giants.'[5]

Their relationship deteriorated irrevocably while Newton was completing the initial edition of the *Principia*, when Hooke and his allies insisted that Hooke had been the first to point out that the elliptical path of the planets round the sun can be described by an inverse-square law of attraction. Determined to retain his priority, Newton wrote angry letters to his colleagues in which he justified his position by protesting that even though the idea might originally have been Hooke's, he – Newton – deserved all the credit for performing the hard mathematical labour. Protesting unconvincingly that he wanted to retreat from the litigious world of natural philosophy, before handing over his manuscript for publication Newton petulantly deleted 'the very distinguished' from his references to Hooke. And it is surely no coincidence that Newton delayed publishing the *Opticks* until the year after Hooke's death.

Newton took critics like Hooke very seriously, not only to establish his priority and protect himself against plagiarism, but also because he recognized the vulnerability of his philosophy. Newton constantly fought to defend his reputation, a task continued by his successors, who worked hard throughout the next three centuries to ward off rival claimants for Newton's unique status. Constantly emended, Newton's texts became riddled with vague conjectures and internal contradictions, as well as statements that would later be denounced as wrong. Yet his propagandists zealously guarded his reputation, so that his ideological role survived unscathed.

As challenges to Newton's hegemony arose, his disciples were forced to protect their master from the attacks of powerful opponents like Leonhard Euler, an eminent member of the Berlin

Academy of Sciences. In the middle of the eighteenth century, he dared to contradict Newton openly. Newton had stated that it is intrinsically impossible to make an achromatic telescope lens, one that does not show coloured fringes round the image. But, Euler provocatively pointed out, Newton was surely wrong, since God had already created His own achromatic lens: the human eye. Irked, the Royal Society deputed the optician John Dollond as spokesman to defend English honour. Dollond argued that Newton's word was tantamount to proof, so that it was 'somewhat strange, that any body now-a-days should attempt to do that, which so long ago has been demonstrated impossible'. Convinced by his own rhetoric that anything 'that great man' had said simply must be true, Dolland failed to repeat Newton's original experiments for almost another ten years. But when he did, he discovered that Euler had been right – Newton was *wrong*! Promptly marketing a revolutionary achromatic telescope based on Euler's discovery, the opportunistic Dollond sat for his portrait smugly clutching a copy of Newton's *Opticks*, a bookmark noting the contested page.[6]

Despite such peripheral snipes, Newton's iconic status continued to grow, even though his philosophical ideas were constantly under attack. Resistance to Newton stemmed not only from scientific and metaphysical reservations, but also from religious and political considerations. It can be mapped socially as well as geographically. In Britain, this opposition lasted particularly long among Tory High Church communities, while in Europe, doubts about adopting a foreign system were often stronger in Catholic countries and absolutist states. By around 1760, the nationalistic implications of this philosophical battle were so widely known that Newton's French and German rivals had even entered patriotic drinking songs:

The atoms of Cartes Sir Isaac destroyed;
Leibnitz pilfer'd our countryman's fluxions;
Newton found out attraction, and prov'd nature's void
Spite of prejudic'd Plenum's constructions.
Gravitation can boast,

In the form of my toast,
More power than all of them knew, Sir.[7]

Textbook writers may have adopted a more restrained style, but they expressed the same sentiments, vigorously denouncing foreign 'romantic systems' of natural philosophy and declaring that 'the *Newtonian* philosophy may indeed be improved, and further advanced; but it can never be overthrown: notwithstanding the efforts of all the *Bernouilli*'s, the *Leibniz*'s, the *Green*'s, the *Berkeley*'s, the *Hutchinson*'s, &c.'. This list, which could easily have been extended, includes some of the Enlightenment's foremost thinkers. Far from being eccentric outsiders, many of Newton's critics were prominent mathematicians and philosophers, whose opinions were highly respected.[8]

Newton's major eighteenth-century opponents – Berkeley, Leibniz and Goethe – levelled different types of criticism that influenced their followers. Berkeley was the most eminent and scholarly mouthpiece for High Church antagonism towards Newton's concept of attraction. Long after his death, Berkeley remained a key source for small religious sects that battered at the increasingly impregnable Newtonian citadel. Leibniz's feud with Newton is fascinating not only because it was so important to both protagonists, but also because it illustrates the complex political ramifications of what might seem an esoteric debate about mathematics. Goethe, now better known for his literary works, formulated a comprehensive optical science that differed fundamentally in its approach from Newton's, and affected the course of German science throughout the nineteenth century. Moreover, he was a founder member of the German cult of genius. Exported to England, Goethe's critiques of Newton reached many artists and poets, thus affecting their antipathy towards science as well as fashioning the meaning of genius itself.

An alternative Principia

A new medical fad swept Britain in 1744. 'Tar water', reported Adam Smith in a letter home from Oxford University, 'is a remedy very much in vogue here at present for almost all diseases. It has perfectly cured me of an inveterate scurvy and shaking in the head.' Five years later, barrels of the new medicine were being exported from Europe all over the world, strongly recommended for potentially fatal diseases such as smallpox and leprosy as well as those common eighteenth-century complaints, gout in the stomach and kidney stones. The inventor of this wonder drug was no fly-by-night opportunist, but a practising physician and distinguished Irish bishop concerned to allay the misery of his famine-stricken parishioners. *Siris*, his long tract extolling the virtues of drinking large volumes of water impregnated with tar, stimulated a virulent pamphlet war between the critics and allies of its august author. However, George Berkeley is now remembered neither as a caring clergyman nor as a dubious doctor, but as one of the century's most eminent philosophers.[9]

About a third of the way through, *Siris* casts off its resemblance to medical promotional material. Laced with forceful denunciations of contemporary immorality, it commutes into a learned diatribe against Newtonian natural philosophy. Berkeley had already roused scholarly controversy by publishing *The Analyst*, a small yet strongly worded book addressed to an unidentified 'infidel mathematician'. Although *The Analyst* was ostensibly designed to undermine the logical foundations of calculus, Berkeley's chief target had been those religious freethinkers who, following Locke and Newton, claimed to discover God by rational thought rather than close perusal of the Bible. Attacking Newton's use of infinitely small quantities in his fluxions, Berkeley argued that mathematics was so plagued with logical errors that by comparison, theological expositions seemed a model of clarity.

Like Berkeley, many Tory High Churchmen advocated

searching for truth through divine revelation, and they opposed arguments based on mathematical abstraction. In *Siris*, his last major publication, Berkeley shifted and broadened his focus. Far from being merely advertising copy for a quack remedy, *Siris* is a weighty tract that explores philosophical and theological objections to Newtonian concepts of matter, space and causality. Berkeley here consolidated some of his earlier objections to Newton's gravitation. To invoke gravity for describing how apples fall to the ground or planets rotate around the sun, is not, Berkeley argued, to provide an explanation. Words such as force, gravity and attraction are useful for carrying out calculations and making predictions, but do not further our understanding.[10]

Siris became a foundational text for natural philosophers disillusioned with Newton's *Principia* and its explanatory claims. 'Attraction', Berkeley protested, 'cannot produce, and in that sense account for, the phenomena, being itself one of the phenomena produced.'[11] Scholars repeatedly worried at the question of whether Newton's power of attraction was a fundamental cause of nature that produced physical effects, or whether it was itself the consequence of some other cause. Many of Newton's early critics derided attraction as 'a late Notion and Assertion in Philosophy, that every thing attracts every thing; which is in effect to say, that nothing attracts any thing'.[12]

This remained a key issue throughout the eighteenth century because it had important theological as well as scientific ramifications. Newton's successors attributed different meanings to attraction, often giving it a physical cause. This meant that debates were often not about what Newton had himself written, but about concepts that later became labelled Newtonian. Some opponents argued that making attraction an inherent property of matter was equivalent to endowing an ordinary stone with holy qualities belonging only to God. For them, positing that bodies could attract each other through empty space opened the door to materialists who would blur the orthodox distinction between active spirit and inert matter, between God and the material world. In principle, by mechanizing the human mind, an extreme materialist could dispense with the concept of an immaterial, immortal soul, thus threatening the basic tenets of Christianity. Critics liberally distributed the pejorative label 'atheist' to warn people away from Newtonians with whom they disagreed.[13]

One influential proponent of Berkeley's critiques was an impoverished curate from Suffolk, William Jones of Nayland. Even in the nineteenth century, some High Church Tories still regarded Jones as their theological inspirer, while readers throughout the country learned about his ideas from books of natural philosophy, and also from browsing through early editions of the *Encyclopaedia Britannica*. As well as publishing acclaimed theological texts and political pamphlets, Jones wrote two long books on natural philosophy. Patronized by the Tory Earl of Bute, who covered his bills at a prestigious London instrument shop, Jones systematically picked apart Newton's concept of gravitational attraction in order to uphold his own insistence that such power could be exerted only by God. Apparently ignoring the friendly concern of his ally the Archbishop of Canterbury that he was becoming 'reputed *an heretic in Philosophy*', Jones boasted that 'Gentleman of the Newtonian side . . . begin to be alarmed about me at Cambridge, & are putting people on their guard.'[14]

Jones cunningly exposed ambiguities in the interpretations of Newton's philosophy by constructing an artificial conversation about attraction conducted between Newton and his chief propagandists. Compiled by juxtaposing ten carefully selected real quotations, Jones's imaginary discussion reduces these grave academics to squabbling children. As a cynical observer commented, he places Newtonians in 'the theatre of a bear-garden, and sets them all a-tilting on a great battle-royal'. Thus when one of them declares that Newton 'considered *attraction* . . . not as a *cause*, but as an *effect*', another contrarily replies, 'Gravity is the most simple of *causes*.' The next one immediately contradicts: for him, attraction means not 'the *cause* of bodies tending toward each other, but barely the *effect*, the *effect itself*'. To which, inevitably, his neighbour responds that God made attraction 'the *first of second causes*' . . . and so on. Rhetorical devices like this could provide effective propaganda weapons for even the most serious scholars.[15]

Jones belonged to a small yet vocal philosophical sect called the Hutchinsonians. A deeply religious and learned man, John Hutchinson had been employed as the Duke of Somerset's steward. As he travelled round the Duke's country estates, he avidly collected fossils, evidence of the earth's divine history. In 1724, the Duke awarded him a pension so that he could retire and compose his riposte to Newton, an ambitious account of the earth's creation that bore a confrontational title: *Moses's Principia*. In a punning *Frontis-Piss*, Hogarth shows these two versions of the *Principia* lying side by side (*Figure 4.1*). While Hutchinson's tome is drenched by the lunar stream that falls inexorably downwards with the force of Newtonian gravity, cabalistic black rats nibble at Newton's book and telescope.[16]

Hutchinson's would-be competitor looked completely unlike Newton's *Principia*. Devoid of mathematical diagrams, and littered with Latin quotations and Hebrew characters, Hutchinson's book focused on Genesis. Hutchinson was horrified that Newton should attribute the power of attraction to matter. For him, it was sacrilegious even to suggest that God, the unapproachable divine essence, could be immanent throughout the corrupt and sordid material world. Newton, he argued, was wrong to study the physical universe in order to learn about God, an approach that veered too close to worshipping Nature itself.

As though preaching from a pulpit, Hutchinson proclaimed that 'the Heathens may take back their Idols of Projection, Attraction, Gravity, Elasticity, &c.'. Instead, philosophers should turn first to the Bible, dictated by God in Hebrew to help people understand the world in which they lived. Like Berkeley, Hutchinson deemed mathematics to be an entirely inappropriate language for discussing God's creation. In common with many of his contemporaries, Hutchinson clung to the pre-Lockean view that words are not arbitrarily attached to objects but somehow in themselves carry a deeper essential meaning. God had – according to Hutchinson – devised Hebrew to make it resonate with multiple meanings relating the spiritual, biblical and physical worlds. This is why Hogarth designed his *Frontis-Piss* as the frontispiece of a book not on natural philosophy, but on the use of points in Hebrew writing.[17]

Believing that human fallibility had led to the scriptures becoming corrupted, Hutchinson devoted years to retrieving and translating the original version, a dedication he shared with other theological experts. Light provides a good example of how Hutchinson scoured the Bible to attack Newton and construct a cosmogony that would conform with his own Trinitarian beliefs. Whereas Newton's followers envisaged light as a hail of little bullets, Hutchinson, Berkeley and other critics insisted that light was a spiritual and not a material medium.

In biblical imagery, God is often portrayed as a fountainhead of knowledge, the source of divine light that flows out to bathe believers in holy illumination. While we may interpret such allusions metaphorically, Hutchinson regarded them literally. His cosmos was a giant machine powered by an aetherial fluid that took three forms – light, fire and spirit, corresponding respectively to God, Jesus and the Holy Ghost. Accusing Newton of limiting God's omnipotence by obliging Him to intervene in the world's operation, Hutchinson envisaged a closed system that had been set in motion by God and would never run down. Unlike Newton's universe with its vast expanses of empty space, Hutchinson's cosmic machine was filled by a circulating spiritual aether that drove along inert matter as it constantly streamed out from the sun to interact with the earth before cycling back again.[18]

Although this alternative *Principia* had only a limited impact during his lifetime, Hutchinson's ideas spread in the second half of the century after his disciples published his complete works. Hutchinson did not always spell out the precise details of how his cosmos might operate, but this vagueness did not obliterate his system's ideological appeal. Hutchinsonianism attracted High Anglicans who felt themselves under threat from both inside and outside the Church. From mid-century, a small clandestine brotherhood flourished that stemmed mainly from Oxford, the country's Tory University, where Jones of Nayland first met Hutchinsonians who later became distinguished public figures. Jones and his influential allies adapted Hutchinson's ideas by eclectically mixing them with concepts and experiments taken from other sources, such as Berkeley, Leibniz and even Newton himself. Yet for polemical purposes, they wrapped up Newtonian natural philosophy with Whig politics and Low Church immorality. Their alternative package was designed to reinforce Trinitarianism and stress human dependence on the word of God.[19]

In addition to this Oxford-based dissidence, some Cambridge graduates also expressed reservations about Newton's approach to nature. Christopher Smart won prizes for his religious poetry, yet his *Jubilate Agno* (Rejoice in the Lamb) was not published until 1939. 'Smart's *Principia*', as this extraordinarily complex hymn has been dubbed, is now fêted as his most splendid creation, yet he composed it while temporarily confined in a lunatic asylum to remedy inconvenient habits such as kneeling in the street to pray. Like Hutchinson before him, Smart was concerned to protect biblical truth from the pretentious claims of natural philosophers: *For I am inquisitive in the Lord, and defend the philosophy of the scripture against vain deceit*. Deliberately making their language resonate with multiple meanings, both authors intertwined the divine, biblical and natural spheres.

In Smart's religious paean, scriptural references jostle with evocative imagery of fire, light and electricity. Parts of it sound like a poetic rendition of Hutchinson's cosmos. Smart left no room for doubt over the correct route to knowledge:

> For Newton nevertheless is more of error than of the truth,
> but I am of the WORD of GOD . . .
> For the MAN in VACUO is a flat conceit of preposterous
> folly.[20]

German rivals

Newton was given a splendid funeral. The pall was carried by six high-ranking aristocrats, the distinguished procession of mourners was headed by a knight of the Bath, and an expensive marble monument commemorated his burial place in one of the most prominent sites in Westminster Abbey (*Figure 2.8*). Eleven years earlier, Newton's opposite number in Germany, the President of the Berlin Academy of Sciences, had received very different treatment. A Scottish visitor reported that Gottfried Leibniz 'was buried in a few days after his decease, more like a robber than, what he really was, the ornament of his country'. Newton's arch-enemy had served the Hanoverian court for over forty years, yet none of his titled colleagues attended the funeral, and he was buried in an unmarked grave.[21]

Leibniz is now world famous as one of Germany's major mathematicians and philosophers, but in the eighteenth century this thumbnail description would have been misleading on three counts: Germany as we know it did not exist, many people had never heard of Leibniz, and he was no university-based academic, but a court employee who travelled throughout Europe on diplomatic missions. Although Leibniz was well known in international scholarly circles, unlike Newton's in England, his was not a household name. The famous astronomer Caroline Herschel, who later emigrated to England, remembered listening as a child to family debates about the relative merits of Newton, Leibniz and Euler, but hers was an unusual experience.[22]

In England, men like Whiston, Cheyne and Desaguliers made versions of Newton's ideas widely available, but in the German-speaking states, Leibniz's ideas remained relatively inaccessible. Leibniz mostly selected erudite journals for publishing his articles, and, like many of his learned contemporaries, often wrote not in German, but in one of the two international languages, French and Latin. After his death, his only major disciple, Christian Wolff, produced dry texts that failed to bring Leibniz anything approaching the national popular acclaim of Newton, his English counterpart.

During Leibniz's lifetime, only a fraction of his myriad writings appeared in print. Volumes are still being added to the latest comprehensive edition of his papers, currently approaching 2 metres of shelf space. These articles often have nothing to do with mathematics or philosophy. As well as dispensing advice on Chinese hexagrams, silver mining, alchemy and fossils, Leibniz dedicated years to compiling genealogical evidence that would justify the dynastic claims of his employer, the Elector of Hanover, who became George I of England.

A voracious reader with a retentive memory, Leibniz's range of knowledge was so impressive that the Elector called him his living dictionary. Yet Leibniz's reputation has, like Newton's, been tailored to fit modern academic categories. Although these two men shared multiple interests in the varied religious, metaphysical and mathematical issues then embraced by natural philosophy, Newton is now celebrated as a scientist, while Leibniz is regarded as a philosopher.

Both were admired as giant intellects by their contemporaries, yet Leibniz and Newton held profoundly different views about how the world operates. While each respected the other's prowess, they disagreed over fundamental issues such as the structure of space and matter, the nature of force, and God's role in the universe. They also became engaged in a bitter controversy over their individual versions of the mathematical technique now called calculus. Their interchanges became increasingly vitriolic. Caroline of Anspach, the most powerful woman in the English Hanoverian court, urged Leibniz towards reconciliation. Writing to him in slightly erratic French, the royal family's language, she preached that 'in this respect, great men resemble women, who never give up their lovers except with the greatest chagrin and mortal fury'.[23]

A few years before Leibniz's death, the focus of his quarrels with Newton shifted on to more explicitly religious and metaphysical territory. Nevertheless, this priority debate dominated the relationships between them and was subsequently vehemently pursued by supporters on both sides. Well into the nineteenth century, Brewster – Newton's partisan biographer – disturbed even his British contemporaries by his exaggerated condemnation of Leibniz. Nowadays, Stephen Hawking is not alone in blurring the historical evidence when he asserts that Newton may have adopted dirty fighting tactics, but was indubitably the first discoverer of calculus.[24]

Who did invent calculus, Newton or Leibniz? This deceptively simple question is intrinsically impossible to answer, since it depends on what one means by invention. It is not even clear that they invented the same thing. Scientific innovations often just cannot be pinned down to a specific time and place: oxygen and Uranus are two classic examples. Moreover, it is often only in retrospect that any one idea appears to be dramatically more important than others circulating at the time. Like his rivalry with Hooke, Newton's conflict with Leibniz now seems uniquely significant, yet it shared many characteristics of the numerous priority debates raging across Europe. Leibniz and Newton themselves both devoted considerable energy to pursuing other similar wrangles, but their dispute over calculus has now become one of the most famous in scientific history. They would both be furious about the Paraguayan postage stamp that shows Newton's head next to Leibniz's integration sign!

Many contestants agreed that Leibniz was the first to publish an outline of differential calculus, and it is his – not Newton's – system that was subsequently developed to form the basis of modern techniques. On the other hand, Newton and his champions claimed that he had originated the basic concept twenty years earlier, and had been openly discussing it in letters with Leibniz and other scholarly colleagues since then. The evidence is hard to assess. Because mathematicians from both camps saw themselves as engaged in warfare, they distorted their opponents' positions, so that the gulf between them now appears wider than it really was: 'Mr Newton', reported one vituperative Scottish acolyte, was 'barbarously . . . and hanoverianlie abused for his principia by a German latelie in a vile consubstantial book of nonsense and ill-nature.'[25]

At a less superficial level, much of the debate was about choosing an appropriate language for mathematizing natural philosophy. In contrast with his private practices, when Newton presented his results publicly he clung to a traditional, geometrical style that established his allegiance with classical philosophers. The continental Leibnizians, on the other hand, boasted about the powerful techniques of manipulation made possible by using symbols, even though these had no basis in physical reality.[26]

In addition, far from being a purely intellectual affair, their rivalry was intimately involved with the political relationships between England and the House of Hanover. Although it was not until 1714 that, as widely anticipated, George of Hanover became King of England, question-marks had hovered over the succession to the throne from the beginning of the century. Leibniz was justifiably concerned about Dutch newspaper reports that his arguments with Newton were seen not as a 'quarrel between Mr Newton and me, but between Germany and England'. Newton, the unsalaried adviser to the English monarchy, was in danger of being eclipsed by his Hanoverian counterpart, and his repeated accusations of plagiarism could only help to discredit Leibniz's standing.

Not all immigrants suffered these problems. George Handel, one of Leibniz's fellow employees at the Hanoverian court, successfully negotiated the move to London by enlisting local patronage and tactfully composing *The Water Music* to herald the new king's barge trip down the Thames. Celebrated as a great English musician, Handel was – like Newton – buried in Westminster Abbey and immortalized in marble by Roubiliac in the act of inspired composition. But Leibniz was instructed to remain behind in Hanover and complete his history of the Guelph family (sadly, he never got further than 1005). Convinced that he was a helpless pawn in an international political contest, Leibniz desperately but unsuccessfully struggled to gain a position at the English court.[27]

Figure 0.1 – Salvador Dalí: *Homage to Newton* (1969). This striking bronze Newton, standing 132 cms tall, is one of three similar Newtons made by Dalí.

Figure 1.1 – Colin Cole: *Design for Baroness Thatcher of Kesteven's coat of arms* (1994). The central shield shows the traditional patriotic emblem of English lions rampant, while the Admiral of the Fleet symbolizes Britain's engagement in the Falklands war under Thatcher's leadership.

Figure 1.2 – Eduardo Paolozzi: *Isaac Newton* (1997). Paolozzi's twelve-foot high bronze statue at the British Library deliberately echoes William Blake's complex picture (*Figure 6.1*).

Figure 1.3 – John A. Houston: *Newton investigating light* (1870).

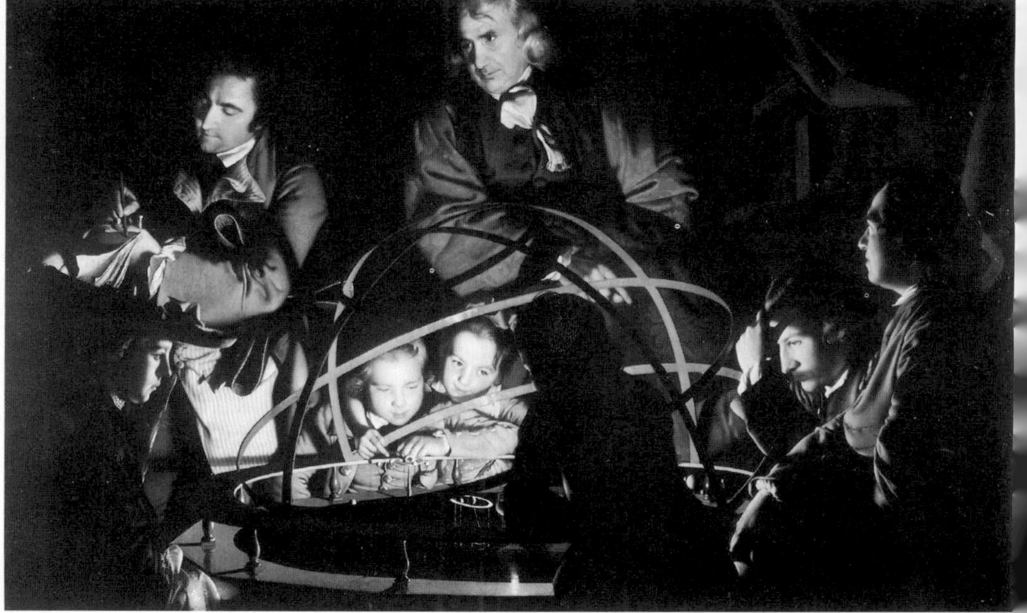

Figure 1.4 – Joseph Wright of Derby: *A Philosopher Giving a Lecture on the Orrery* (1766). Many of Wright's paintings illustrate the growing enthusiasm for natural philosophy during the eighteenth century, when orreries became a familiar symbol of Newtonian cosmology and the hierarchical Georgian social order.

Figure 1.5 – Frontispiece of the first English translation of Newton's *Principia* by Andrew Motte (1729). It quotes a version of some lines from Halley's long poem to Newton, especially composed for the *Principia*: 'The hidden secrets of heaven and the immobile order of things lie open, with Mathematics pushing the cloud away. Now the keenness of a sublime Intellect has allowed us to penetrate the dwellings of the Gods and to scale the heights of Heaven.'

Figure 1.6 – Frontispiece of Voltaire's *Elémens de la philosophie de Newton* (1738). Divine light is transmitted by Newton to be reflected from Truth's mirror, here held by Mme du Châtelet, down on to the Enlightenment's most distinguished scribe, Voltaire.

Figure 2.1 and *2.2* – Godfrey Kneller: *Isaac Newton* (1689) (*left*) and *Isaac Newton* (1702) (*right*). Kneller painted two versions of the 1689 picture (*left*). The half-length from which this engraving was made (via a photograph) belonged to Newton and was inherited by the Earl of Portsmouth; the other version, a three-quarter length, was probably owned by Caroline of Ansbach, and it shows a Greek book of Daniel, a reference to Newton's preoccupation with biblical prophecy.

Figure 2.3 – John Vanderbank: *Isaac Newton* (1725).

Figure 2.4 – Engraving of Newton based on Godfrey Kneller's *1720* portrait.

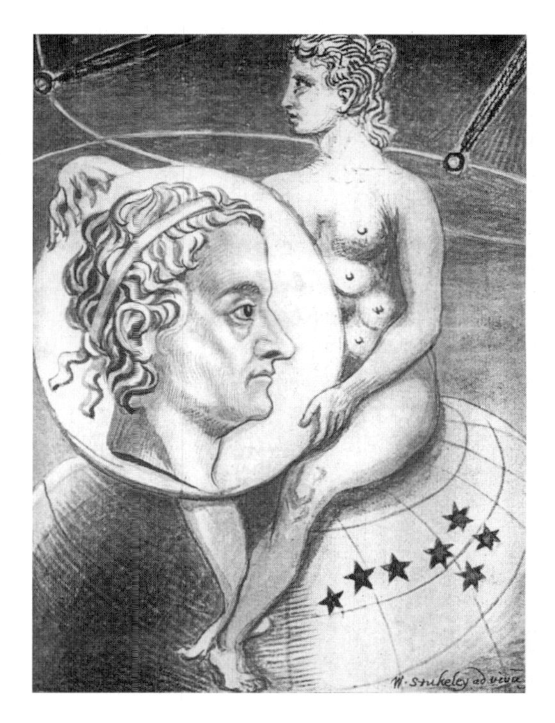

Figure 2.5 – William Stukeley: pen and wash drawing of Newton, c.1720. Reflecting his own antiquarian interests, Stukeley has drawn Newton in profile, as though he were a Roman emperor on a medal.

Figure 2.6 – George Bickham: *Isaac Newton* (1787). This engraving was first published in 1732, and then reissued in 1787 by Bickham's son with accompanying verses from a Stowe estate poem.

Figure 2.7 – William Hogarth: *A Performance of 'The Indian Emperor or The Conquest of Mexico by the Spaniards'* (1732).

Figure 2.8 – Anthony Pugin and Thomas Rowlandson: *Entrance into the Choir of Westminster Abbey* (1812). Newton's monument, designed by William Kent and sculpted by Michael Rysbrack, lies to th left of the entrance, and Lord Stanhope's monument lies to the right. This picture conveys a good impression of how these statues appeared when they were first installed.

ROMERO,
RUDY

Wed Sep 14 2016

Newton : the making of genius /

33029047692199

Hold note:

*

*

*

*

*

*

*

*

Terminal # 260

09/13/2016

Artículos

TITULO Newton : the making of

COD BARRAS 33029047692199

DEVOLVER 10-04-16

TITULO The Da Vinci code / Dan

COD BARRAS 3 3029 04311 5674

DEVOLVER 10-04-16

TITULO Einstein's brainchild :

COD BARRAS 33029042723874

DEVOLVER 10-04-16

TITULO Dorland's illustrated

COD BARRAS 3 3029066438771

DEVOLVER 10-04-16

¡Muchas gracías por visitar

la biblioteca!

Sacramento Public Library

www.saclibrary.org

Terminal # 250

Figure 2.9 – Louis François Roubiliac: Newton's statue at Trinity College, Cambridge (1755).

Figure 2.10 – Giovanni Battista Pittoni, with Domenico and Giuseppe Valeriani: *An Allegorical Monument to Sir Isaac Newton* (1727 – 30). There were two versions of this picture, but only this one was engraved. The other one shows Newton's monument in Westminster Abbey, probably thanks to Conduitt's influence.

Figure 2.11 – Jakob Houbraken after Kneller: allegorical portrait of Newton.

Figure 2.12 – R. Page: engraving of Newton in the 1818 edition of the *London Encyclopaedia*.

Figure 3.1 – Frontispiece of John Colson's *The method of fluxions and infinite series* (1736). In this idyllic scene of rural England, the precision of Newtonian techniques and English gun manufacture have surpassed the calculations of Greek philosophers, here shown anachronistically wielding quill pens. The Greek motto at the bottom means 'the common things in an unusual way, the unusual things in a common way.'

The philosophical systems proposed by Newton and Leibniz were related to the social systems within which they had been formulated. Leibniz can be regarded as the product of a vanishing medieval and scholastic world view that was in the process of being displaced by a new social order, of which – with hindsight – Newton now appears as an early representative. Many implications of the political and theological differences between Leibniz and Newton were hammered out in a three-cornered exchange of letters between Queen Caroline, Leibniz, and Newton's apologist Samuel Clarke, a London cleric associated with the court.

Protagonists on both sides often used the metaphor of a clock to portray the conflicting accounts presented by Leibniz and Newton of how God superintends the universe. On Newton's model, God is constantly active throughout the cosmos, and intermittently exerts His supreme power to intervene and alter the laws of nature. Leibniz was scathing about this view: 'Nay, the machine of God's making, is so imperfect, according to these gentlemen; that he is obliged to clean it now and then . . . and even to mend it, as a clockmaker mends his work.' Surely, he protested, God was no sloppy mechanic, but a skilled craftsman who could initially wind up His clock to run perfectly throughout eternity. According to Newton, God had created independent, individual particles that, as they travelled through empty space, constantly interacted with each other and formed new associations. In contrast, Leibniz maintained that God had established a harmonious universe completely filled by inherently active entities called monads. Although they operated independently, and no longer needed God's direct control, Leibniz's monads had been in a sense pre-programmed so that they worked together to fulfil His plans.[28]

Models of the universe provided blueprints for how society should be organized. Germany consisted of small, separate feudal states with individual rulers, but before the Union with Scotland in 1707, England was a single country. Her Glorious Revolution of 1688, when the people rejected traditional monarchy to give an elected parliament a new and uniquely powerful role, is often heralded as the precursor of the American and French Revolutions. Brought up in this rapidly changing national context, Newton introduced a revolutionary cosmology that involved the shifting interactions between independent, mobile particles. Taken up and reinterpreted by his successors, Newton's model of the physical world helped to promote the development of democratic societies based on the principle of individual equality.

While Clarke argued – on Newton's behalf – that it was a king's responsibility to govern through active intervention in his people's activities, Leibniz contended that a good ruler makes sure his subjects are sufficiently well educated to run their own affairs. As a seasoned diplomat, Leibniz was skilled at devising encoded messages, and his metaphorical terminology ties together his cosmological and political views. 'All intelligences or souls capable of reflection [*rulers*]', he wrote to the uncle of the French foreign minister, 'have important privileges, which exempt them from the revolutions of the bodies [*the people*] . . . Together they form the republic of the universe, whose Monarch [*Louis XIV of France and Leopold I of Hanover*] is God.' The small German principalities resembled monads, each represented by a central court, whose balanced political relationships with each other paralleled the pre-established divine harmony of the universe.[29]

In England, as Newton and his contemporaries manoeuvred to consolidate their positions under the shifting political regime, they hotly debated issues such as free will, miracles, and the divine appointment of kings. To his consternation, Leibniz recognized that opposition to his philosophical ideas was being related to conflicts between English groups with differing religious and political alignments. He 'would not have imagined', he protested, 'that the spirit of faction would go so far as to spread even into the mathematical sciences'. For Newton and his allies, it became advantageous to suggest that their English opponents were spreading ideas similar to those of Leibniz, and hence denounce them by association with this German intruder.[30]

Reciprocally, during the first half of the eighteenth century, natural philosophers in the German-speaking lands were unenthusiastic about Newton's innovations. They were far more interested in developing the ideas suggested by Leibniz, a scholar whose discussions of philosophical questions resonated with local political and theological concerns. Wolff, the most influential German metaphysician of this period, relied heavily on Leibnizian concepts, and it is tempting to caricature the debates of this period as a head-on confrontation between the Wolffians and the Newtonians. The situation was, in actuality, far more complex. Wolff became embroiled with religious opponents, and most German scholars were not reading Newton's books, but modified versions of his philosophy coming out of Leiden.

In 1746, Pierre Maupertuis became President of the Berlin Academy of Sciences. A decade earlier, he had been one of Newton's most energetic propagandists in France, and his German accession was converted into Newton's symbolic overthrow of Leibniz. In his best-selling international textbook on natural philosophy, Euler described with relish how Wolff and his sectarian supporters had cursed the entire Academy as they ungraciously conceded defeat. But, like Maupertuis, Euler was no pure disciple of Newton. As well as touting his anti-Newtonian achromatic lens, he had already relegated Newton's geometrical language to the past by rewriting the *Principia* with his version of Leibniz's calculus. Like other so-called Newtonians, both Maupertuis and Euler formulated their own distinctive philosophical positions by drawing on a variety of sources, which included Leibniz and Descartes as well as Newton. By the time that Newtonianism became the accepted party line among German academics, its precepts differed substantially from those proposed by Newton himself.[31]

Leibniz's ideas also initially fared better than Newton's in Russia, a country hosting many German scholars and where Leibniz himself had acted as personal adviser to Peter the Great. Naval and military men rapidly saw the advantages of adopting Newton's optical instruments and mathematical techniques, but his philosophy found few supporters. Even one of Newton's admirers, a professor at the Academy of Sciences in St Petersburg, declared that 'I have always desired those excellent celestial instruments whose invention is the glory of Newton' but dismissed gravity as 'an occult quality from the old Aristotelian school which converts healthy people to madness'. Peter envisioned Russia as a centralized, coordinated bureaucracy to be governed in a hierarchical, ordered manner. Like the components of a smoothly running clock, his courtiers would work together to serve the interests of the Tsar and the state. Leibniz's cosmology of preordained harmonic integration matched this model of society far more closely than Newton's view of independent individuals interacting freely with one another.[32]

In Britain, an articulate, well-informed and politically effective public voice developed far earlier than in other European countries. As individual achievement became increasingly valued, newspapers, theatres and other intellectual institutions became potent mediators between the ruler and the people. Newton's philosophical system matched the political one. Its strong endorsement of personal autonomy within a close-knit community underpinned his own promotion as a celebrity throughout British society.

But Leibniz's fame in the German-speaking lands was restricted to narrow academic circles, especially after the Berlin Academy officially pledged allegiance to Newton. German scholars have noted with regret that it was a French Swiss editor who produced the first so-called complete works, fifty years after Leibniz's death. Further collections were published sporadically, but it was only at the beginning of the twentieth century that his status was reappraised by several international scholars. Working independently, they hailed Leibniz as a neglected genius who had already formulated the most recent discoveries in mathematical logic. Leibniz's philosophy, declared Bertrand Russell, has 'been hitherto universally and completely misunderstood. This is to be accounted for partly by his sheer intellectual greatness, partly by the ignorance of editors, partly . . . [by] the vastness of the enterprise he undertook . . .'[33]

Towards the end of the eighteenth century, long after Leibniz had died, the town of Hanover did eventually erect a temple dedicated to their local genius, but by then it was two other German writers – Immanuel Kant and Johann Goethe – who were becoming acclaimed as national geniuses.

Kant's philosophical texts have the reputation of being particularly abstruse, yet although he lived roughly a century later, it is Kant and not Leibniz who became Germany's genius of rationality, the nation's closest equivalent to Newton. Although adopting the life-style of a reclusive academic, Kant – unlike Leibniz – propelled himself into the public spotlight, intervening in contemporary political debates and campaigning for freedom of the press. With its ringing call '*Sapere aude!* Have courage to make use of your *own* understanding!', his prize-winning essay 'What is enlightenment?' provided a manifesto for progressive rational thought.[34] Kant held mixed feelings about Newton's scientific ideas, but in many respects they became cultural counterparts. We remember Kant primarily as a philosopher, but during the nineteenth century he was also an iconic figurehead for German scientists.

Kant embodied the power of pure reason, the subject of his most famous book. Like all cult heroes, his reputation has been embroidered with colourful anecdotes, such as his identical daily walks timed with such precision that the townspeople used him to set their clocks by. Corroborating legendary tales of his abstemiousness and eccentricity, his portraits and death-mask depict a spare figure and a gaunt face. After his death in 1804, phrenologists declared that Kant's status as a genius was confirmed by his skull, which revealed a brain possessing overdeveloped intellectual bumps but only minuscule ones of vanity, and absolutely none of the features associated with sexual desire. A century later, one school of German psychiatrists investigating insanity invented a new medical term, schizophrenia, to describe the mental tendencies of methodical, logical scholars with delicate bodies. Such geniuses were labelled the 'Kant type', a group that also included Newton.[35]

Romantic rainbows

Did Kant regard himself as a genius? Despite the phrenological evidence attesting to his lack of vanity, on a modern meaning of genius the answer is certainly yes. But his own comments are inconsistent. Near the end of his life, in a book based on some early lectures, Kant pronounced that 'The genius is a man not only of wide range of mind but also of intensive intellectual greatness, who is epoch-making in everything he undertakes (like Newton and Leibniz).' Conveniently bracketing his two great mentors, Kant's fuzzy definition sounds like a self-description, but in his major writings he introduced a new and very different concept of genius that was to prove enormously influential.[36]

Driving a wedge between artistic creativity and rational thinking, Kant taught that genius is a talent for producing an original piece of work that operates completely independently of any rules. Bizarre though it might seem to us, he declared that Newton was definitely *not* a genius: whereas Newton could clearly demonstrate every step on the path he had taken to reach his theory of gravity, original writers of genius like Homer were incapable of describing their thought processes, because they did not themselves know how they had obtained their ideas. Following his own sharp cleavage of the mental world, Kant would certainly not have classified himself among the geniuses; he belonged to the 'far superior' group of 'great men' who relied on logical thought and argument to make great discoveries and to teach others.[37]

Kant's intellectual sneer at geniuses was partly directed at some of his own contemporaries, the young writers of the *Sturm und Drang* (Storm and Stress) movement. Flourishing briefly during the 1770s and 1780s, this group articulated fresh concepts of creativity and national identity that stemmed partly from earlier British books on literary and aesthetic originality, notably by Edward Young. Reacting against the earlier optimistic faith in progress expressed particularly forcefully by Leibniz, they rejected the power of reason and launched a cult of original genius, whose adherents focused on the imaginative, mysterious aspects of human nature. Overturning traditional models of authorship, they viewed literary works as the products of neither a skilled craftsman nor a muse listening to divine dictation, but of a genius driven from within by an internal source of inspiration.[38]

Kant singled out for comment the best-selling author Christoph Wieland, whose prose translations helped to convert Shakespeare into an icon of creativity for aspiring German authors. But by far the most famous advocate of original genius was Goethe, now seen as the founding father of the Romantic movement. In 1744, embroidering his own experience of rejection by a woman, he published his hugely successful novel about Werther, a hyperemotional young man who kills himself for unrequited love. In contrast with the measured orderliness of Kant's solitary walks, Goethe's own love-strewn life helped to originate a new stereotype of the original, Romantic genius who is governed only by his own creativity as he plunges from one extreme of feeling to another. Goethe's poems and plays catered to the widespread wish for an indigenous middle-class literature, written in German, that would displace the Francophonic culture of the nobility and unite the separate German-speaking states into a single nation.

Throughout his life, Goethe constantly revised his ideas about genius and himself came to personify German genius. If Kant is the German Newton, Goethe came to represent the cultural equivalent of Shakespeare, an original genius who had created the nation's greatest literature. But – in Germany at least – he is also remembered for his scientific activities, partly because chauvinistic nineteenth-century biologists claimed that he was a precursor of Charles Darwin. An enthusiastic collector who owned 18,000 rock samples, Goethe engaged in international discussions about human anatomy, botany and geology, as well as writing extensively on the physiology and physics of colour.

Unlike Kant, Goethe believed that science should not be conducted exclusively by logical, systematic thinkers such as Newton. Although he admired Newton enormously, he wanted to eliminate the split being forced between abstract scientific analyses and everyday experiences of life. For him, the emotional intensity and heightened awareness of a literary genius were also valuable attributes for creating a more humane type of scientific knowledge. So his optical experiments included studying the effects of gazing at a woman's bright clothes or a snow-covered mountainside, while his novel *Elective Affinities* modelled marital exchanges on molecular transformations.[39]

Of all the scientific controversies in which Goethe became involved, the one that aroused most hostility – but also most enthusiasm – was his sustained attack on Newton's optical theories.[40] In his most famous scientific book, *Colour Theory*, Goethe attacked Newton in terms so abusive that even he later regretted them, ingenuously blaming his vehemence on the Napoleonic wars. For modern art historians, Goethe's *Colour Theory* has become a classic study of colour whose status parallels that of Newton's *Opticks*, a founding text for scientists. It influenced, for example, the Munich Blue Rider painters of the early twentieth century, who repeated Goethe's experiments of gazing at scenery through a prism to see coloured fringes, and also revived his polar and spiritual theories. Scientists, on the other hand, now dismiss Goethe's ideas with scorn, even though he had many enthusiastic followers during the nineteenth century.

Contradicting Newton's claim that white light is a mixture of many colours, Goethe maintained that colour arises from mixing white light with darkness, its polar opposite, so that coloured fringes arise at a black edge against a pale background. He preached a subjective approach towards scientific experimentation, deliberately including rather than excluding the observer's own reactions. Campaigning to found a new human-based science of chromatics, Goethe wrote extensively about the structure of the eye, the perception of different colour qualities, and the relevance of his theories to art.

Newton – so runs the idealized account – set a prism in the path of a ray of light pouring through a small hole in his window shutters, and examined the coloured spectrum cast on to the opposite wall (*Figure 1.3*). In a sense, he converted his own eye into a scientific instrument that could objectively analyse an image from which it was completely detached. In contrast, Goethe used not a wall but his own retina as a screen, looking through a prism held directly in front of his eye to experience different colours radiating out in opposite directions. For Goethe, the observer cannot be separated from the image under scrutiny.

Rainbows came to symbolize the differences between Goethe and Newton. Laden with religious symbolism, rainbows were a favourite topic for aspiring young artists who wished to demonstrate their philosophical allegiance as well as their technical expertise. In her allegorical self-portrait, Goethe's friend Angelika Kauffmann showed herself in the act of painting a prominent rainbow across the sky. Goethe designed his own publicity material for a pack of instructional playing cards: like a Masonic emblem, his own eye glares out from the centre of the woodcut, projecting rays of light that stream beyond the arching rainbow.[41]

Newton, strongly influenced by Greek ideas of cosmic musical harmony, had asserted that there are seven colours in the rainbow, although rainbows had often previously been shown with only three bands. Seven has, of course, always been a significant number in magic and astrology as well as music, and one might say that Newton invented the colour indigo to make his rainbow conform with his theoretical beliefs. Newton's rainbow was central to his optical thought, and it first reached artists and literary writers through poetry, especially James Thomson's *The Seasons*; this was, for instance, one of the artist Joseph Mallord William Turner's favourite books. Like other eighteenth-century poets, Thomson translated Newton's terse experimental accounts into verbal pictures that lyrically glorified the rainbow, God's natural equivalent in the sky of Newton's man-made spectrum on his study wall:

> . . . *First, the flaming* Red
> *Sprung vivid forth; the tawny* Orange *next;*
> *And next delicious* Yellow*; by whose side*

Fell the kind beams of all-refreshing Green.
Then the pure Blue, *that swells autumnal Skies*
Æthereal play'd; and then, of sadder Hue,
Emerg'd the deepened Indico, *as when*
The heavy-skirted Evening droops with Frost.
While the last Gleamings of refracted Light
Dy'd in the fainting Violet *away.*[42]

Goethe's house in Weimar is still decorated with a defiantly anti-Newtonian rainbow, which has only three clear colours and is inverted so that blue lies at the top and red at the bottom. Even some English and French painters deliberately depicted rainbows displaying their disaffection with Newton's theories.[43] Ironically, like Newton's, Goethe's thought was profoundly influenced by his alchemical studies. At the heart of Goethe's anti-Newtonian onslaught lay his concept of 'augmentation', which describes how a semi-opaque medium enables blue and yellow, the two extreme tones, to produce red, the highest, noble hue.

When Goethe's *Colour Theory* came out in 1810, it prompted mixed reactions. English readers regarded the very notion of an anti-Newtonian science as a contradiction in terms, and were incensed by the audacity of this challenge to Newtonian ideology from a foreign poet. When the translation appeared in 1840, some particularly offensive sections were prudently omitted. Even so, Brewster, ever prejudiced, fumed that this 'bold, though unbidden minstrel' had 'assailed the mild precepts of Newton with . . . all the sophistries of German metaphysics'. Even such a marvellous poet was, in his eyes, unworthy to covet 'the diadem of *Newton*'.[44]

The German scientific response was not so dismissive. Goethe's contemporaries shared his emphasis on polarities, which resonated with their own investigations into the role of opposites in magnetic, electrical and chemical activity. The enormously influential philosopher Georg Hegel, for instance, supported Goethe against Newton in his *Encyclopædia*, while physiologists immediately started to credit Goethe's contributions to their studies of perception. Some of Europe's leading young intellectuals, including Coleridge and Shelley, were trying to integrate living beings within their theories of the physical universe. They studied phenomena that affected people as well as inanimate matter, such as animal magnetism and galvanic electricity. Just as Goethe had used his own eye as a recording instrument, experimenters tried to cure themselves magnetically and made their own bodies part of electric circuits. Non-Newtonian science featured significantly in the lives of England's Romantic poets.

Complementing these scientific investigations of an anti-Newtonian approach, two artists – Philipp Runge in Germany and Turner in England – used paint to explore Goethe's insistence on colour's psychological impact and moral symbolism. Runge epitomized the tragic artistic genius: this idealistic visionary even conformed to stereotype by dying young of tuberculosis, an illness then widely believed to be linked with genius.[45] Runge's early death meant that he left his ambitious schemes uncompleted, but his ideas about colour harmony strongly influenced later artists and are still cited today. Like Goethe, with whom he corresponded extensively, Runge imbued his eerily luminescent imagery with symbolism based on mystical philosophies as well as colour theories. Thus his three primary hues – blue, yellow and red – signify respectively the Father, the Holy Ghost and the Son, but also indicate different times of day and carry various gender associations.

Thirty years later, when the English translation brought Goethe to English artists, Turner was struck by the ways in which Goethe's list of polar opposites meshed with his own interest in light and shade, and he painted a complementary pair of episodes from Noah's flood. One is an evening scene of despair and punishment painted in dark brooding blues, while its companion, dominated by a bright yellow swirling centre, is called *Light and Colour (Goethe's Theory) – the Morning after the Deluge – Moses Writing the Book of Genesis*. Although prompted by Goethe, these paintings demonstrate Turner's own unshaken belief in sunlight as the source of colours. Similarly to Runge, Turner was intrigued by Goethe's arguments, but by no means convinced that all of them were right. Perhaps turning to his volume of Thomson's poetry, he painted rainbows whose diffused hues deviated from conventional representations but conformed to Newton's order.

Rainbows provided a potent emblem for English Romantic writers to accuse Newton of destroying nature's beauty. Like Kant, they wanted to drive a wedge between scientific knowledge of the world and artistic, intuitive representations. Keats's warnings about eliminating a rainbow's mysteriousness by dissecting its components have become a particularly famous example:

> *Do not all charms fly*
> *At the mere touch of cold philosophy?*
> *There was an awful rainbow once in heaven:*
> *We know her woof, her texture; she is given*
> *In the dull catalogue of common things.*
> *Conquer all mysteries by rule and line,*
> *Empty the haunted air, and gnomed mine –*
> *Philosophy will clip an Angel's wings . . .*
> *Unweave a rainbow, as it erewhile made*
> *The tender-person'd Lamia melt into a shade.*[46]

Keats had evidently been thinking about rainbow imagery for some time. Three years earlier, drinking at a dinner hosted by Benjamin Haydon, he had agreed with the 'excessively merry and witty, . . . and tipsey' essayist Charles Lamb that Newton 'had destroyed all the Poetry of the rainbow, by reducing it to a prism'. The revellers' toast to 'Newton's health, and confusion to mathematics' has often been cited as the slogan of a supposed Romantic anti-Newtonian movement. However, even taking this cheerful mockery at face value, there was no simple party line of opposition to Newton or to science. Blake is frequently held up as the epitome of Romantic scientific hostility, yet even his rainbows are firmly Newtonian ones. At this celebrated dinner, Newton only entered the conversation in the first place because Haydon was allocating him a favoured spot in his still unfinished painting, *Christ's Entry into Jerusalem*, which hung above the drinkers. In what would become his most famous picture, Haydon deliberately contrasted a sneering, atheistic Voltaire with the man 'whose intellect in its range almost reached the outermost circle of the influence of Divinity ... Newton, who was a Believer'.[47]

Wordsworth obligingly laughed along with the other guests, but spent over thirty years working on his lines about Newton's statue in Trinity College chapel. Coleridge, Wordsworth's poetic mentor, echoed Goethe's conviction that Newton was wrong to believe in the possibility of objective, detached observation: '*Mind*, in his system, is always *passive*, – a lazy *Looker-on* on an external world. If the mind be *not passive*, if it be indeed made in God's Image ... there is ground for suspicion that any system built on the passiveness of the mind must be false, as a system.'[48]

Wordsworth tried out different ways of rendering poetically how he himself, like Goethe and the German self-experimenters, was involved in the act of perception. In *The Prelude*, he imagined looking down from his room in St John's towards Newton's statue in the antechapel of Trinity, the adjacent College. He rejected attempts that incorporated himself, deleting phrases such as 'And from my pillow' or 'pressing on my sight'. Instead, Wordsworth eventually decided to retain the plain disengaged 'I' as solitary beholder. His final version evokes the concept of Newtonian detached vision:

> *... I could behold*
> *The Antichapel where the Statue stood*
> *Of Newton with his prism, & silent face,*
> *The marble index of a Mind for ever*
> *Voyaging throu' strange seas of Thought, alone.*[49]

5

FRANCE

From the fact that the action and reaction of opposing powers is always equal, the greatest efforts of the Goddess of Reason against Christianity were made in France.

Joseph de Maistre, *Considerations on France* (1797)

Promoting Newton in Europe could be a hazardous business. In Italy, Francesco Algarotti's *Newton for the Ladies* was placed on the Vatican's Index of banned books, and Paolo Frisi risked censure by contrasting Galileo's persecution with Newton's celebration.[1] Frisi was one of the numerous European philosophers who held England up as the land of liberty in order to criticize other regimes, so that debates about natural philosophy were fraught with political and religious implications. In order to convince French readers that Newton should supplant Descartes, Voltaire was obliged to publish in Amsterdam and London, while Jean Delisle de Sales was imprisoned for the materialist views he expressed in his *Philosophy of Nature*.

Like other best-sellers that circulated secretly in Enlightenment France, Delisle de Sales's *Philosophy of Nature* was published abroad, and appeared in small pocket-size volumes that were convenient for clandestine reading. In his controversial book, Delisle de Sales included a short play about Newton in which he satirized contemporary discussions about faith and reason. For the dramatic opening scene, Newton (who in real life never even travelled as far as Oxford) is perched by the coast in Senegal, taking time off from checking his calculations of the tides to contemplate the grandeur of nature (*Figure 5.1*). The plot hinges on Newton being a vegetarian who intervenes in various dialogues between a merman, an oyster and a native African about whether or not they should eat each other. Newton, here set up as an icon of rationality, concludes that only the African is worth teaching, since his perception of God as a cockchafer does at least indicate that he can acquire a human soul.[2]

For modern readers, the humour of a play that could be broadly summarized by the Cartesian quip 'I eat, therefore I think' appears somewhat strained. However, Delisle de Sales was not the only author who imaginatively incorporated Newton within settings that strike us as bizarre, but which made serious comments on profound philosophical issues. For instance, in 1748, Denis Diderot, one of the two major editors of the *Encyclopédie*, published an erotic novel, *The Indiscreet Jewels* (perhaps he regretted this venture the following year when he was arrested for producing seditious literature). French aristocrats snapped up copies of what turned out to be Diderot's most popular work, which metaphorically vaunted female sexuality as well as Enlightenment rationality. Eager purchasers evidently relished the pornographic wit articulated by the 'jewels' (sexual organs) of gossiping women, but they also appreciated Diderot's subversive, thinly veiled references to prominent figures in the French establishment, including the King. In one chapter, Newton features in a sultan's dream about the phallic rise (and collapse) of science. Another scene is set at the Academy of Banzo, where Circino, the Newtonian philosopher of attraction, engages his Cartesian rival in a debating duel, significantly conducted under the light of the full moon.[3]

Finding Newton in under-the-counter literature comes as rather a shock, but by the middle of the eighteenth century he had become a cultural figurehead well outside academic circles. Delisle de Sales's spell in prison indicates how dangerous materialist philosophies were perceived to be. On the other hand, that Diderot should use Newton to spice up soft pornography illustrates to what extent philosophical controversy was a popular topic of conversation. Newton's rise to iconic status depended not only on the direct propagation of his ideas in elite intellectual texts, but also on his frequent appearances in a huge range of poems and pictures, books and buildings. Using extravagant language, French poets celebrated Newton as a 'tow'ring genius . . . Sagacious! comprehensive! and sublime!'. More pithily, a facetious society gossip reported that 'One may see Lawyers forsake the Bar to busy themselves in the study of Attraction, and Divines neglect their Theological exercises for its sake.'[4]

By the end of the century, this English hero had become the French God of Reason. This was far from being a straightforward process. At the Paris Academy of Sciences, Frenchmen argued about the validity of some aspects of Newton's theories right through the eighteenth century. For one thing, chauvinistic interests dominated the acceptance of philosophical systems. In a letter to a French colleague, an English mathematician regretted the universal 'Tyranny of Prejudice' affecting 'even the most zealous and industrious Searchers after Truth . . . We have our Newton, the Germans their Leibniz, and you your Descartes.'[5]

Long after Newton's optical ideas had been broadly accepted, eminent natural philosophers rejected his concept of gravitational attraction, and many of them never agreed with various aspects of his work. But different theories were gradually cut and pasted together to create new syntheses in a Lego-like approach to constructing models of the universe. By the end of the century, French people who thought of themselves as Newtonian were in reality imbibing and disseminating an international blend of Newtonian, Cartesian and Leibnizian approaches to nature.

Reactions to Newton's ideas were strongly affected by religious debates. Most controversially, the Baron d'Holbach relied on Newtonian principles in his *System of Nature* of 1770. Perceived as hugely threatening, this notorious yet influential manifesto of Enlightenment materialism was not translated into English for almost a century. D'Holbach horrified many readers (including those who had not opened the book) by denying the existence of the human soul. He argued that people are made solely from matter, so their behaviour must be entirely governed by mechanical laws. Claiming that the basic urge driving human behaviour is self-preservation, d'Holbach reformulated Newton's first law of motion: 'Newton calls it force of inertia, moralists have called it in man self-love . . . This gravitation is thus a necessary disposition in man and in all beings, who . . . tend to persevere in the existence they have received, as long as nothing disturbs the order of their machine or its primitive tendency.'[6]

Often pointing to d'Holbach's book, devout Catholics bracketed Newton, Descartes and Leibniz together, fearing that *any* natural philosophy posed a threat to religious belief. Even towards the end of the century, several contradictory positions thrived at the same time, because Newton could be reinterpreted to suit different theological interpretations. Most Jesuits denounced Newton as an atheist and a materialist who had constructed a Godless universe. But other experts enlisted Newton on the opposite side, insisting that his orderly cosmos proved the existence of a divine architect. This was, for instance, the position that the elderly Voltaire came to defend.[7]

But despite these scientific and religious debates, Newton was converted into an iconic figurehead. As part of their rationalizing propaganda, Enlightenment philosophers held him up as a shining exemplar of the invaluable contributions that scientific knowledge could make to social progress. The *Encyclopédie*'s introduction, often taken as the defining manifesto of Enlightenment thought, gave pride of place to Newton, celebrating him as the century's leading genius who, supplanting Descartes, had 'appeared at last, and gave philosophy a form which apparently it is to keep'.[8] Because gravity could be praised as a democratic force that affected everyone equally, Newton became a hero for Revolutionary *citoyens* (citizens). But at the same time, he remained '*le chevalier anglais* (the English knight)', leader of a new intellectual aristocracy.

Renowned for his mathematical approach to nature and his work on optics, Newton personified the two major preoccupations of French Enlightenment philosophers – reason and vision. The very term Enlightenment conjured up the strong bonds between seeing and knowing, between lucidity and rationality, an ocular paradigm that ruled particularly strongly in France. French writers, architects and artists played with multiply punning imagery of mental and optical illumination, of cosmological and terrestrial order. Rays from the Sun King had shone over but also controlled his subjects, while the Masonic all-seeing eye so prevalent in Revolutionary iconography indicated God's omniscience. Similarly, Newton had wielded a prism to analyse the light of God, and had focused his mind to deduce how gravity bound His universe together.[9]

Pursuing this optical metaphor, the *Encyclopédie* placed 'the philosopher at a vantage point, so to speak, high above this vast labyrinth [of human knowledge], where he can perceive the principal sciences and arts simultaneously'. Poets expressed the same image more lyrically. As they hymned Newton's celestial vision and power, they often gave him the piercing gaze of a high-flying eagle, traditional symbol of a genius's sight, or transformed him into a spirit roaming the heavens. French poetry of this period does not enjoy a good reputation. Modern readers do not relish long poems whose dedications start:

> *O SHINING SPIRIT! . . .*
> *To thee, whose eye sounded the depths of the universe,*
> *Great shade of Newton, I address my verses!*

Nevertheless, such tributes played a vital role in consolidating Newton's fame and confirming the semi-divine status that he was acquiring.[10]

Well before the end of the century, it had become perfectly consistent to praise Newton as a genius and yet disagree profoundly with aspects of his philosophy. How could an English natural philosopher, whose theories were never fully accepted, be lauded as a semi-divine genius under a French egalitarian regime? One way of resolving such apparent paradoxes is to consider that Newton became fêted not so much as an individual in his own right, but rather as a transcendent entity, an abstract idealization, so that pictures, poems and buildings represented not Newton himself, but the concepts of reason, order and genius that he came to personify.

Voltaire's presence looms so strongly that he is often credited with having brought Newton to the French people almost single-handedly, but Newton's renown travelled along numerous other routes. His elevation to international glory was unique. Our understanding of why and how this happened is greatly enriched by looking at some of the commemorative statues, poems and buildings that reached people who held no deep knowledge of gravity or optics, but whose sentiments affected the conduct of science.

Displacing Descartes

One of the more contrived jokes circulating among French wits concerned a Newtonian and a Cartesian who had a fight. Because the Newtonian lacked any repulsive force, he fell to the ground when the Cartesian's fist was attracted to his centre instead of being deflected in a circle.[11] But even sophisticated commentators reduced the complex French debates about natural philosophy into a power struggle between two adversaries.

Like Newton, Descartes was a national figurehead who carried great symbolic importance even for those who had only a scanty knowledge of his theories. For years both heroes coexisted – sometimes joined by Leibniz and Euler – as icons of modern progress. But as science and philosophy gradually became separate academic disciplines, Newton became heralded as a scientific founding father who had superseded his French predecessor. Descartes, on the other hand, was placed in the philosophical canon, so that although his research in mathematics and experimental physics came to seem less significant, he retained his heroic status.

Shortly after Newton's funeral, Voltaire provided an instantly quotable contrast:

> A Frenchman arriving in London finds things very different, in natural science as in everything else. He has left the world full, he finds it empty. In Paris they see the universe as composed of vortices of subtle matter, in London they see nothing of the kind . . . For your Cartesians everything is moved by an impulsion you don't really understand, for Mr Newton it is by gravitation . . .

Writing in political exile, Voltaire was keen to make France seem reactionary in comparison with libertarian England, and he emphasized that Newton was the greatest man who had ever lived because he governed by truth rather than violence.[12]

In his funeral eulogy to Newton, the distinguished French philosopher Bernard de la Fontenelle offended sensibilities on both sides of the Channel by bracketing him with Descartes. Both of them, he told the horrified Paris Academy of Sciences, were geniuses. Immediately published in French and English, Fontenelle's tribute to an English hero caused a minor European sensation: Charles-Louis Montesquieu even demanded that copies be couriered out as gifts for his hosts in Vienna. Conduitt, who had provided Fontenelle with many biographical details, later complained that he had not done 'justice to that great man who had eclipsed the glory of their hero Descartes', but victory was neither as swift nor as total as Conduitt claimed to believe.[13] Even Fontenelle warned his audience not to succumb to the temptation of believing in attraction, and although at mid-century the *Encyclopédie* insisted that Cartesianism had been banished from France, this bold assertion was itself an exaggerated advertisement for Newtonian ideas.

Far from being a palace coup, Newton's succession was slow, patchy and complicated. In the 1670s, his early work on optics had initially met with a cool reception. Intrigued yet sceptical researchers found they could not replicate his results, so that Newton gained the reputation in French academic circles of being a brilliant mathematician but a ham-fisted experimenter. Convincing confirmation only came some forty years later, when

Desaguliers redesigned many of Newton's original experiments. Fortunately for Newton's international renown, natural philosophers converged on London in 1715, ostensibly to celebrate the Hanoverian King's accession to the throne, but more probably to witness a total solar eclipse. Fanning foreign enthusiasm with diplomatic gifts and dinner invitations, Newton and his allies ensured that the visitors carried experimental instructions back home, and also encouraged the publication of a handsome Parisian edition of the *Opticks*.

Meanwhile, the *Principia* had not gained many European adherents. Among the few readers competent to judge it, most agreed that while Newton had produced an inspired mathematical hypothesis, it bore little relationship to physical reality. But during the 1730s, dogmatic resistance started to soften. It was a Jesuit priest, a self-taught disciple of Descartes called Nicolas Malebranche, who first convinced loyal Cartesians to incorporate some elements of Newton's ideas into their own world view. Although Malebranche and his admirers never did relinquish their swirling vortices of particles, they were won over to Newton's laws of gravitation. This partial acceptance paved the way for the more militant Newtonian revisionists who succeeded them.[14]

But outside narrow academic circles, an ambitious young mathematician called Pierre Maupertuis played a far more dramatic role in persuading the French nation to embrace English cosmology. Already the author of an explicitly Newtonian book on astronomy, Maupertuis was an inspired self-publicist whose rise to fame had as much to do with his skill at convincing different factions of his expertise as with his experimental results. Maupertuis benefited from the revival of an old debate, one which provided the grounds for Voltaire to quip that 'In Paris you see the earth shaped like a melon, in London it is flattened on two sides.'[15] Cartesians envisaged the earth as being slightly compressed round its middle, while according to Newton, it should be flattened at the poles.

The French Academy of Sciences decided to resolve the issue by sending teams to Peru and Lapland during the 1730s. On paper, comparing measurements taken as near as possible to the Equator and the North Pole sounds like a simple way of distinguishing conclusively between these competing theoretical positions, but the real-life situation was, inevitably, far more complex. The adventurous explorers were relatively inexperienced, cosseted young gentlemen who suddenly had to contend with extremes of temperature and terrain, while in Paris, experts argued from the comfort of their drawing rooms about the best type of instrument to use, and the validity of the measurements. Far from being neutral arbitrators, members of both camps had made up their minds in advance what the answer would be: personal reputations and national pride were at stake in this supposedly scientific contest.

Although the equatorial expedition lasted ten years, Maupertuis and his colleagues in Lapland completed their work quickly and, in 1737, triumphantly returned to Paris clothed in furs. The results were ambiguous, but Maupertuis insisted that, though contested, these measurements vindicated his own faith in Newtonian gravitation. The earth's flattened shape was, he declared, now established beyond all reasonable doubt. Fanning public sympathy with colourful newspaper tales of his team's arduous battles against fog, mosquitoes and freezing cold, Maupertuis even published anonymous satires lambasting his opponents. The same poets who enthused that Newton had torn away the Cartesian bandages of error also hymned Maupertuis for his 'astonishing audacity' as he tackled 'mountains of ice' and 'braved torrents hung suspended in space'. Only 'the finest motives inflamed the hearts' of the plucky explorers. Backed by Newtonian propagandists, Maupertuis became an exemplary French hero whose courageous venture entitled him (and France) to some of Newton's glory.[16]

Two of Maupertuis's earliest converts were Voltaire and his lover Émilie du Châtelet. She condemned Cartesianism as 'a house collapsing into ruins, propped up on every side . . . I think it would be prudent to leave.' As they set about persuading the nation to desert with them, Voltaire's reverence for Newton came to verge on obsession. He scattered praise throughout his books and letters, while dinner-party guests were obliged to admire his bust of 'the greatest genius that ever existed: if all the geniuses of the world were assembled, he should lead the band'.[17] Promptly taking advantage of the surge in interest stimulated by Maupertuis's Lapland success, in 1738 Voltaire published his *Elements of Newton's Philosophy*. Although simplified and distorted, this illustrated account of optics and gravity became one of the major sources of French familiarity with Newton's innovations.

Although Voltaire was happy to receive the credit, this man of literature did recognize that he was far less knowledgeable about mathematics and natural philosophy than his erudite lover. In the frontispiece of his book, du Châtelet appears as the goddess of truth reflecting an attenuated beam of divine Newtonian light down on to Voltaire, who is wearing a classical robe and a poet's laurel wreath (*Figure 1.6*). As a preface for his Newtonian primer, Voltaire composed a long poem that hymns Newton, but opens by addressing du Châtelet as his 'great and powerful genius, France's Minerva'.[18]

Du Châtelet studied intensively and made original contributions to a range of topics. In particular, her translation of Newton's *Principia*, with its substantial scholarly commentaries, was admired for its clarity and its superiority to the English version. Despite her domestic duties, du Châtelet did manage to complete the manuscript, but after she died in childbirth, it was another ten years before a full version was published. Voltaire frequently extolled her achievements; they would both be gratified to know that hers remains the only complete French translation.[19]

Faced with the problem of overcoming resistance to female scholarship, du Châtelet invited the country's leading intellectuals to participate in discussions at her estate at Cirey. The behaviour of one of du Châtelet's guests infuriated her. She had learned Italian expressly to engage in philosophical repartee with Algarotti, and he was probably delighted to visit Cirey rather than accept his other invitation – to join Maupertuis's Lapland expedition.

Nevertheless, in his *Newtonianism for the Ladies*, Algarotti made it clear that du Châtelet was the model for his fictional Marchioness, and so cruelly parodied her penchant for fine clothes and jewels. His frontispiece shows Algarotti as a slender elegant man ('a swan', sneered Voltaire) accompanying an unflattering portrait of a rather substantial du Châtelet, while his dialogue parodies her as a flirtatious intellectual lightweight. For instance, he put into her mouth such Newtonian absurdities as 'after eight Days Absence, Love becomes sixty-four times less than it was the first Day, and according to this Progression it must soon be entirely obliterated . . .'[20]

But another Cirey visitor, the fashionable artist Maurice Quentin de Latour, presented a very different view of a Newtonian woman. At the 1753 Louvre exhibition, which was – as usual – packed by crowds from all the social classes, he displayed eighteen new portraits of Enlightenment men and women. Several journalists singled out for comment his vibrant pastel showing Elizabeth Ferrand, a fashionable Parisian hostess, studying a book on Newton (*Figure 5.2*).[21] Unlike the male-dominated club culture of England, in France women's *salons* were recognized as social centres where serious discussions took place. Despite being a semi-invalid, Ferrand exerted considerable behind-the-scenes influence in Enlightenment France, and regularly engaged in debates about current political and scientific affairs with her eminent friends. Although some colleagues sniped in the press that she was mean and humourless, everyone paid tribute to her mathematical prowess. One man who publicly acknowledged his intellectual debt to Ferrand was the famous philosopher Étienne Bonnot de Condillac. The heart-felt sincerity of his overt tribute in his *Treatise on the Sensations* pays her a far truer compliment than does Voltaire's overblown poetic praise of du Châtelet.[22]

Latour often depicted his female subjects engaged in intellectual pursuits (although when he painted Madame de Pompadour surrounded by books and globes, hostile critics accused him of detracting from her femininity). His picture of Ferrand, briefly interrupted from her Newtonian studies, reflects the barriers du Châtelet encountered when she tried to escape from her prescribed role as a spectator, and become an active participant in scientific life. Using the luminous pastel colours for which he was renowned, Latour shows Ferrand temporarily distracted from her peaceful perusal of Newton's philosophy. Gazing out from a state of absorption, whether lost in reading, grief or contemplation, was a favourite theme for eighteenth-century artists, and moralizing poets preached the innocent delights of combining usefulness with pleasure. Latour's scene suggests that Newtonian texts had entered the canon of morally improving books for women, so that Ferrand apparently embodies the Enlightenment ideal of a feeling heart guided by reason.[23]

This portrait is, however, shot with ambiguities. Ferrand's clothes – her lace-lappeted cap and pink negligée trimmed with blue – indicate that the artistic viewer has been invited into a private domestic space. Her quizzical gaze and the coquettish angling of her fingers, which recall François Boucher's intimate portrait of his wife seductively lounging in her *boudoir*, belie Ferrand's studious pose, and suggest that she is theatrically engaging with the spectator. The book's careful lettering deceptively suggests fidelity to an actual original, but this large volume with its marginal annotations more closely resembles the Bible than any available edition of Newton's philosophy. Newton was already starting to assume the role of an idealized figure who bore scriptural authority in the secular realm of philosophy.[24]

By this time, access to Newton's ideas had become far easier. For those who could read neither English nor Latin, Newton himself had intervened to ensure that a fine new translation of the *Opticks* was produced, and du Châtelet's French *Principia* did eventually appear. At the Sorbonne, Diderot's friend Pierre Sigorgne (who was at one stage imprisoned for passing on subversive political poetry) was lecturing on Newton's natural philosophy to his students, and in 1747 he published his own influential textbook on Newton. In addition, French translations of explanatory guides like Pemberton's became available. The first volumes of the pro-Newtonian *Encyclopédie* were starting to appear, and the small simplified books by Voltaire and Algarotti did, despite their inadequacies, advertise Newton to non-academic French readers. In Paris, fashionable audiences were flocking to the entertaining demonstration lectures given by Jean-Antoine Nollet. Inspired by a visit to Desaguliers, Nollet included many Newtonian experiments, which reached a wide readership through his enormously successful six-volume illustrated lecture course.

But not everyone approved of this spread of Newton's natural philosophy. Even at the heart of France's scientific community, powerful anti-Newtonian factions operated well into the second half of the century. When Halley's comet reappeared in 1759, Newtonian astronomers – notably Joseph-Jérôme Lalande – had to work hard to convince their sceptical colleagues that their English methods yielded the most accurate predictions. Although Newton's allies had long viewed his cometary analyses as particularly telling arguments against Cartesian vortices, opponents continued to insist that French cometographers had been successfully perfecting their own techniques over the previous sixty years without recourse to Newton. Many of their objections were based on flaws in Newton's own calculations, yet just as Maupertuis managed to convert his ambiguous results into definitive proof of polar flattening, so Lalande presented the return of the comet as a clinching success story for Newtonian celestial mechanics.[25]

The sustained Jesuit resistance to Newton had a very significant effect throughout Catholic Europe because the order controlled many of the best educational institutions. Jesuits clung to Cartesianism, partly because it drew a sharp distinction between the soul and the body. By choosing to give Newton's philosophy a materialist interpretation, they could strengthen their own position by denouncing his adherents as atheists. Jesuit book reviewers sneered at enthusiasts who had deserted Descartes to leap on the Newtonian bandwagon: 'At last Mr Voltaire speaks, and immediately Newton becomes famous or is on the way to becoming so; all Paris reverberates with Newton, all Paris stutters Newton, all Paris studies and learns Newton.'[26]

Selective translation provided one powerful strategy for subverting Newton's influence. Substantial portions are missing from the French version of Algarotti's *Newton for the Ladies*, and the Jesuit editor provided a preface criticizing Newton and praising Descartes. But head-on confrontation was the most obvious tactic. Because they wrote in Latin, Jesuits who wanted to attack Newton could reach an international audience. Several of them wrote anti-Newtonian poems, sometimes with extensive footnotes explaining that Newton's philosophy led to atheism. One of the most widely read was the Cardinal de Polignac's *Anti-Lucretius* of 1747, which was promptly translated from Latin into French. Dedicated to promoting 'Religion and Virtue', this long poem (483 pages in the English version) is resolutely Cartesian:

> *Nor does great NEWTON's famous system stand,*
> *On one compact foundation, simply plann'd . . .*
> *Reflect how vainly is that Art employ'd,*
> *Which founds a stately fabrick on a Void:*
> *Confess the fair result of sober thought,*
> *WHO BUILDS ON VACUUM, MERELY BUILDS ON NOUGHT.*[27]

As late as the 1770s, investigators all over France participated in a protracted debate about gravity. Extraordinary as it now seems, a scientific journal published details about research carried out in the Savoy mountains claiming to show that Newton was wrong – objects get *heavier* rather than lighter the further away they are from the centre of the earth. It was several years before it became clear that the original experiments had never taken place and that the article's authors were fictional characters. That this fraud should be so successfully perpetrated illustrates for how long powerful anti-Newtonian lobbies survived in France.[28]

Because Newton and Descartes were both important symbolic icons, they could be attacked or praised by protagonists defending a wide range of positions. To say one belonged to a Newtonian or a Cartesian sect was socially significant, but by no means implied total intellectual commitment. There was no more uniformity amongst the neo-Cartesian camp than among the Newtonians they opposed, and many writers sought to combine their approaches. One common solution was simply to ignore the philosophical problems raised by attraction, and instead regard Newton's cosmology as a useful model for making accurate quantitative predictions about the world. By the time of the Revolution, an ostensibly united scientific facade concealed huge rifts between factions holding very different opinions about aims and methodology. Nevertheless, Newton's supporters had won the ideological battle, and – as in England many years earlier – it became impossible to shake Newtonian orthodoxy. Thus the Academy of Sciences automatically rejected the optical theories proposed by the physician Jean-Paul Marat because they contradicted Newton's. Marat did, however, enjoy a successful political career before his untimely end in a bath.

Anglomania

Just as Voltaire exaggerated the contrast between Newtonian London and Cartesian Paris, caricaturists on both sides of the Channel delighted in portraying national stereotypes. In a British engraving of 1768, a clumsy, overweight Englishman is unsuccessfully trying to learn how to dance under the disdainful eyes of his slim, foppish teacher, who despite being unmistakably French, is wearing a hair ribbon labelled Newton. This foreigner might, the artist implies, be forcing his employer into an inferior role, but even he must pay tribute to England's intellectual hero. A decade earlier, a Frenchman had coined the term Anglomania to describe the French fad for all things English – including Newton – that continued up to the Revolution. Initial resistance evolved into deliberate imitation, and English ways of life profoundly affected the geographical, intellectual and political landscapes of France during the second half of the century. 'Albion', enthused one poet, was the home of Newton. 'Oh how the universe should envy that island!'[29]

In England, the widespread dissemination of portraits, busts and statues made a vital contribution to Newton's conversion into an English hero. However, in France visual imagery played a different role. Pictures of Newton himself enjoyed only a limited circulation. His admirers – men like Buffon and d'Alembert – did seek inspiration by hanging Newton's portrait on their study walls, and several engravings and medals were reproduced in books or could be bought separately. But many of these had been derived from other copies, and came to bear little resemblance to the original, often looking characteristically French. Particularly when Newton appeared as one in a series of great men, he tended to lose his personal identity. Moreover, there were virtually no French statues or busts of him. After the Revolution Newton had become a French icon, but the state encouraged paintings and sculptures that would celebrate national achievement and democratic ideals. The nave of Westminster Abbey contained private memorials to famous individuals, but the Panthéon, Paris's secular cathedral, housed representations of abstract quantities like Reason and Architecture.[30]

Because Newton was symbolically important, images claiming to show his actual appearance were less significant than pictures and monuments intended to convey the spirit, as it were, of Newton and his philosophy. One such building that survives is the Temple des Philosophes at Ermenonville, now about an hour's drive from Paris. Its history illustrates some of the ways in which Newton became absorbed into French life and politics. This classical Temple with six Tuscan columns nestles among trees on a slope overlooking a lake, and the current neglect of this prefabricated ruin heightens its melancholic atmosphere. Newton's column is placed next to Descartes's, while the other four geniuses are Voltaire, Penn, Montesquieu and Rousseau.

Like many wealthy French landowners, during the 1760s and 1770s the Marquis de Girardin redesigned his estate at Ermenonville in the English allegorical style made so famous by Stowe.

The tight mathematical arrangements of formal French gardens gave way to rolling parklands that were carefully crafted to appear natural. Dotted with classical statues and Masonic grottoes, this sculpted scenery articulated in material form the sentimental vision of nature that artists and writers were adopting from English pictures and poetry. The meticulously planned groves and monuments laid out by Girardin and other landed gentry reflected how English customs were affecting not only how French people interacted with their environment, but also how they thought about death and commemoration.[31]

Girardin frequently indulged in 'a calm and tranquil morning *à l'anglaise*' perusing Newton and Milton, and spent tranquil evenings with English visitors, exchanging poetic snippets from the Newtonian poet Thomson and his French imitators. Among the rustic monuments scattered around his grounds, Girardin erected an obelisk to four pastoral poets, including James Thomson. Like the pyramid on Newton's tomb, the obelisk's shape symbolized eternal fame, while the tribute to Thomson (inscribed in English) summoned up Newton's astronomy and optics:

Like the circling sun; his
Warm genius
Coloured and vivified every
Season of the year.

Ermenonville became a pilgrimage site after Jean-Jacques Rousseau was buried there in 1778. Although the obelisk with its Newtonian inscription was demolished by Revolutionary troops as a symbol of foreign oppression, swarms of Sunday visitors still enjoyed wandering round the Temple des Philosophes while an orchestra played under a nearby beech tree. But this elegant building, with its splendid views over the lake, had originally been designed as a tribute to genius and a comment on the finite duration of time. Based on the temple at Tivoli, a favourite spot for young men on their Grand Tour, the rotunda is supported by six pillars dedicated, Girardin explained, to 'those privileged Geniuses who appear briefly to honour their country & illuminate their peers'. Now overgrown with nettles, two adjacent columns are inscribed to Descartes and Newton, here commemorated for his work on light, not gravity. In this prefabricated ruin, the seventh column is intentionally broken, while three others lie nearby on the ground, awaiting the arrival of another comparable genius.[32]

Whether in private contemplation or public salon readings, ardent Anglophiles like Girardin particularly favoured two English poems that eulogized Newton and his ideas: Edward Young's *Night Thoughts* and Thomson's *The Seasons*. Young inspired astronomical poetry that envisaged Newton's soul ecstatically soaring through the heavens 'on the wings of Genius, holding Urania's dividers in his hand'.[33] But it was Thomson in particular who became a cult hero, and whose references to Newton were often quoted in English as well as French. Thomson had completed his *The Seasons* by 1730, and thirty years later the first French translation stimulated French versions. Compared with Thomson's lyrical Georgic hymns to nature's beauty, much of this derivative French poetry seems rather leaden, using rhyming couplets to intone moral homilies about the virtues of hard work.

Newtonian poetry became politicized in the hands of Thomson's French emulators. One of the best known was Jean-François de Saint-Lambert, such a favourite visitor at Cirey that he fathered du Châtelet's child. In his French *Seasons*, Saint-Lambert urged his aristocratic readers to care for the peasants who tended their property; unlike Thomson, he focused on the industrial value of minerals rather than their beautiful spectral colours. Scientific creativity, he stressed, stems from social need, while happiness comes not from whiling away the long hours of winter darkness at balls and concerts, but in quiet solitude learning how to 'enlighten oneself between Locke and Newton!'[34] Another of Thomson's imitators obliquely criticized royal oppression by praising Newton as a scientific liberator, a 'daring eagle' who had been brave enough to 'look the blazing King of the Skies in the face' and break 'the chains of ancient ignorance'.[35]

On the other hand, not all French people were enthusiastic

about this cultural invasion from across the Channel. It was not only natural philosophers who resented Newton for supplanting their own national hero. One (fictional) society hostess complained about a faddish young visitor who 'has the *Anglo-manie* to a great degree. He unfortunately spent fourteen days in London, and speaks of it incessantly; is always boasting of the learning and genius of the English . . . He keeps English horses, reads the English newspapers, makes his morning visits with boots and spurs, drinks tea twice a day, and thinks himself as wise as Newton or Locke.'[36] In the middle of the century, some writers still heralded Descartes as Europe's intellectual saviour: 'without that great Frenchman, whom we must regard as the Founder of sound Philosophy, Great Britain would still be groaning under the tyranny of the Aristotelians'.[37]

Anne-Robert-Jacques Turgot, the influential public administrator who had been one of Sigorgne's early Newtonian pupils at the Sorbonne, resolved the conflict by giving these two national heroes equal status in the empire of science:

> Some people have made it their business to sacrifice the reputation of Descartes to that of Newton. They have imitated those Romans who, when one emperor succeeded another, simply knocked off the head of the first and substituted it for that of the second. But in the Temple of Glory there is room for all great geniuses.

Turgot insisted that scientific advance depends on a continuous succession of great men who open up new avenues of research to be explored by lesser minds. He was articulating the faith in rational progress professed by the rationalizing French philosophers of the Enlightenment. For them, nature was a more important source of knowledge than God, and they enrolled Newton into a new intellectual elite intended to replace the older aristocratic and clerical hierarchies.[38]

Turgot's vision of a tightly knit scientific community still survives. When art historians look back at the work of the sculptor Robert-Guillaume Dardel, they automatically bracket together two of his statues, one of Descartes and one of Newton. Although Dardel is not famous now, he moved in prestigious circles, and examining his sculptures provides insights into the bonds between scientific and political interests. Because these two pieces were produced eleven years apart, on either side of the Revolution, they carried very different political messages and should not be considered as a pair.

Dardel first exhibited a terracotta model of 'Descartes piercing the dark clouds of ignorance' in 1782. Now in London's Wallace Collection, Dardel's Descartes is himself realistically represented in seventeenth-century clothes, but he is sitting astride an amorphous shape resembling rounded rocks. These are the metaphorical clouds of ignorance symbolically being penetrated by the sun's rays. On one interpretation, this Descartes is an emblem of scientific achievement, but to portray him in this way was also to make a patriotic declaration. In the same exhibition, Dardel showed a deliberately flattering sculpture of his patron, the Prince of Condé, pursuing his retreating enemies. As France's wealthy aristocrats were becoming increasingly concerned about the nation's decline, sculptors were producing expensive monuments that consoled them by nostalgically recalling the glories of the country's vanished past. In England, Wedgwood was mass-marketing cheap models and medallions of Enlightenment intellectuals from all over Europe. But in France, only rich customers could afford the Sèvres replicas of these statues, which had been designed to create a pantheon of specifically French achievement. Dardel was subsequently commissioned to produce a full-scale version of his Descartes, which, paralleling his statues of famous military men, publicly proclaimed the greatness of France's intellectual heritage.[39]

In 1793, Dardel decided to sculpt Newton. Unfortunately, his terracotta model probably no longer exists, but much can be deduced from the informative title: 'Newton discovers and displays [female] Truth, who in one hand holds a prism to indicate the theory of colours, and in the other a magnetized circle to designate his system of attraction'. Whereas at the time of his

Descartes pieces Dardel had been employed by an aristocrat, he now played a prominent role in post-Revolutionary artistic politics. Dardel was renowned for his use of allegory, and since the three other sculptures he exhibited that year were laden with republican symbolism, his Newton certainly carried concealed meanings. At this time, Dardel's major patron was Jacques-Louis David, whose splendid double portrait of the chemist Antoine Lavoisier and his wife resonates with allusions to how scientific expertise could contribute to social improvement.[40]

In Dardel's vanished sculpture, Truth's prism symbolized not only Newton's optics, but also the rays of Enlightenment. The 'magnetized circle' remains mysterious, but may well indicate that Dardel knew about recent French experiments extending Newton's law of gravitational attraction to cover magnetism.[41] Although natural philosophy was a male preserve, Truth and Reason were, as in Voltaire's frontispiece, always personified as female (*Figure 1.6*); they were, moreover, Revolutionary icons in contemporary polemical art. Dardel's Newton was rooted in these allegorical associations between scientific knowledge and Revolutionary progress. By exhibiting Newton, Dardel was making a political statement about French ideals of social reform through scientific research and technological innovation. Newton had become an iconic figurehead, a God of Reason who featured prominently in polemical tracts on equality, stability and progress.

Astronomical order and gravitational equality

The year before the French Revolution, Louis Fontanes published a long didactic poem on astronomy that launched him on his doubly successful career as a poet and a politician. Inspired by a recent trip to England, Fontanes took his cue from Young's *Night Thoughts* to stress the insignificance of human achievement and the complexity of God's design. Newton, he declared (in what to modern readers seem rather agonizing rhyming couplets), was 'the most worthy of God's interpreters', who had lifted one corner of heaven's veil to reveal 'which laws govern the motion of all the planets'.[42]

Fontanes belonged to a new generation of French writers who had been taught Newtonian physics and were keen to celebrate modern ideas by incorporating them within lyrical yet factually informative poetry. Poetry was an important genre of Newtonian education, one which explained scientific concepts but also delivered political, religious and moral messages. Poets often produced versified versions of the Enlightenment slogan that reason would dispel the mistaken beliefs and superstitions of earlier, more ignorant ages. Furthermore, in the charged atmosphere of the post-Revolutionary decades, many of them stressed that Newton, 'that great legislator of the worlds and Heavens', had organized and stabilized the universe. Thanks to Newton, they proclaimed, 'error was dethroned, and his dividers brought order and simplicity'.[43]

United by their praise for Newton, these poems conveyed contrasting opinions about science, knowledge and progress, and were designed for different audiences. Nevertheless, their common emphasis on Newton's uniquely powerful role reinforced his fame among wide groups of readers who lacked any formal scientific education. Some of these literary tributes read like breathless sports commentaries about intellectual athletes pounding towards truth at the top of Mount Olympus. But whatever their poetic qualities, as well as publicizing Newton, they also helped to consolidate the concept of a scientific community transcending national boundaries, and to reinforce the increasing power of scientific institutions after the Revolution.[44]

In their prefaces and footnotes, these didactic poets frequently acknowledged their debt to Joseph-Jérôme de Lalande and Jean-Sylvain Bailly, two eminent astronomers who had published influential textbooks supporting Newtonian over Cartesian cosmology. It was Lalande, so devoted that he made a point of visiting Westminster Abbey to bow at Newton's tomb, who had successfully converted the 1759 comet into a vindication of Newtonian prediction. Renowned for his dedication to the public understanding of science, he swallowed dried spiders to discourage arachnophobia, and embarked on dangerous balloon flights. Less spectacularly, but perhaps more importantly, he encouraged astronomical poetry and did much to promote the spread of Newtonian ideas throughout the French educational system. Bailly, who established his own laboratory in the Louvre and was particularly well known for his detailed history of Newtonian astronomy, later became Paris's first elected mayor, when he was satirized in political caricatures as an absent-minded star-gazer tumbling into a well. After he was guillotined in 1793, it was Lalande who read his funeral eulogy.

Lalande and Bailly inspired poetic interpretations of Newtonian astronomy in which the theme of scientific progress, so central to Revolutionary ideology, was particularly important. It was usually given a strong nationalistic twist. Newton appeared in an ancestry of great men stretching back in time to Aristotle, Kepler and – above all – Descartes, the vital predecessor without whom (they claimed) even Newton would not have been able to formulate his theories. The lineage of intellectual aristocrats also pointed forward to Newton's successors, French Newtonians such as Maupertuis and Lalande.

Lalande sponsored a didactic history of astronomy designed for schoolboys. This long chronological tour ranging from the Chaldeans to the present included detailed tables of astronomical distances, as well as technical footnotes about celestial mechanics, and was delivered in resounding rhyming couplets. Newton appeared as the sole discoverer of nature's hidden secrets, while Maupertuis and his colleagues comprised specifically French role models of bravery ('French audacity conceived, dared and achieved a hundred prodigious feats undreamt of elsewhere . . .').[45]

By the time of the Revolution, the members of the Paris Academy all regarded Newton with reverence, yet several different scientific outlooks sheltered beneath this Newtonian umbrella. Lalande championed what we might term the hard physics school of thought, whereas his critics disapproved of analytical techniques and deterministic cosmologies. In line with Jean-Jacques Rousseau's sentimental pastoralism, they sought to develop a more holistic philosophy. It was this approach that inspired *Letters to Sophie*, an enormously successful educational poem.

In strong contrast with Lalande's adventurous hymns to progress, *Letters to Sophie* delivered moral lessons suitable for girls, teaching them that the goal of science was to learn about God by uncovering nature's harmonious structure. Letters from an avuncular tutor urged Sophie and her friends to abandon their jewels and lace and turn instead to their globes. Expected to grapple neither with mathematics nor Arctic ice, these female readers were gently persuaded to admire the immortal Newton for revealing the beauty of God's creation, and to marvel how 'at the genius's voice, the stars make a sweet harmony heard'.[46]

A third prevalent view at the Academy was that science's major aim should be to improve social welfare through invention. This became a favourite theme for progressive poets. One of the most famous was André Chénier, who had worked at the London Embassy before the Revolution, and after his return belonged – like the sculptor Dardel – to David's intellectual circle. Educated at one of the first schools to include Newtonian science in its regular curriculum, Chénier became a leading advocate of political scientific poetry. Criticizing stodgy versifiers, Chénier campaigned for a new poetic style that would capture modern scientific thought while avoiding pedantry. In *Invention*, his most renowned doctrinaire poem, Chénier fêted Newton as the crowning glory of a cumulative chain of achievement. Newton, he wrote, was made to 'talk in the language of the Gods' as Urania (goddess of astronomy) taught Calliope (muse of poetry) to string 'her golden lyre to a nobler pitch'.

A political moderate, Chénier used his rhymed version of Bailly's Newtonian astronomy to draw forceful analogies between cosmological and political order. He symbolized the universe's suns as kings ruling over the planets that circled around them, weighed down by the yoke of intransigent laws 'whose sacred, essential, inflexible power makes them all yearn towards an invisible centre'.[47] After Louis XVI was executed, it became dangerous to express such monarchist sentiments. André Chénier was imprisoned and then guillotined, despite the efforts of his younger brother, Marie-Josèphe, who was a leading Jacobin politician, to prevent his death.

Already an established playwright, Marie-Josèphe Chénier wrote stirring hymns praising Revolutionary virtues. In a long poem celebrating the power of reason, he also paid tribute to Newton, albeit far more vaguely than had his elder brother. In French, Newton conveniently rhymes – well, almost – with the words for reason, Cicero, Cato and Plato. This enabled Marie-Josèphe Chénier to list Newton as one among many intellectual heroes who had been inspired by 'exquisite reason'. His verses may not be great poetry, but they do illustrate how Newton had entered an international, timeless pantheon of great men.[48]

By the late eighteenth century, Newton had become the God of Reason, an emblematic hero symbolizing stability, equality and rational progress for people who had only a hazy notion of his achievements. For those who escaped André Chénier's fate, the stability of Newton's universe provided an obvious metaphorical contrast to the Reign of Terror. In lengthy allegorical epics, poets compared the periods of mistaken beliefs that had interrupted scientific progress before Newton with the wars, decadence and famines that afflicted the development of human civilization. Comets, which were the subject of intense debate and apprehension, provided particularly potent symbols of political chaos and divine retribution. As a scientific genius, Newton was becoming hailed as a secular prophet. 'Nations! rest assured,' wrote one poet; Newton has declared that comets 'will one day reanimate the ageing Sun . . . and I believe in his genius.'[49]

Newtonian physics also provided a political metaphor for equality. In one anti-monarchic adaptation of Thomson's *The Seasons*, Newton was praised for ordering the universe so that it 'resembles those wise kingdoms, where one law is enough to control all men. Attraction: that is the law of the universe.' Nature now provided an egalitarian political model, in which every individual was governed by the same law of attraction. Unlike the packed, swirling universe of Cartesian vortices, the Newtonian void was populated by separate bodies that – as another poet put it – 'one & the Other attract each other by the same bonds, while at the same time, & following the same laws, they weigh towards a common centre'.[50]

While poets hymned Newton's genius verbally, some architects were expressing the concepts of Newtonian order and celestial infinity in the visual language of buildings. A small but important group of innovators, now often called the Revolutionary architects, conceived their buildings as poems communicating abstract ideas through a universal language of shapes. Intended to embody Republican virtues, their geometrical plans broke with existing conventions, and were designed to evoke appropriate sensations in the spectator. Henceforth, they declared, appearance was to be strictly related to function. Although their more visionary schemes were, like those of eminent modern architects, never constructed, their ideas and ideals radically affected the appearance of Paris as well as towns throughout Europe.[51]

One of the most influential members of this group was Étienne-Louis Boullée, an architectural teacher for over fifty years, who sought to regenerate architecture through geometry, and to revive aesthetic and religious values. In his buildings, Boullée used pure mathematical forms to symbolize the transcendent order of the Newtonian universe and the harmony between people and their environment. Working with Bailly's *Astronomy* on his bookshelf and a portrait of Newton on his study wall, in 1784–5 Boullée designed two versions of a cenotaph to Newton, both essentially gigantic spheres (*Figure 5.3*). These Platonic shapes immediately summoned up the immensity of the Newtonian universe, as well as the earth's symmetrical shape before its Newtonian flattening. More obliquely, they referred to the recent French invention of air balloons, spherical emblems of a democratic science that would enable ordinary people to soar above the heads of their social superiors.

Addressing Newton as a divine being, Boullée explained his

inspiration: 'While by spreading your illumination and the sublimity of your genius, you have determined the shape of the earth, I have conceived the project of enclosing you within your discovery.' Initially, he planned to illuminate the interior of his sepulchre with a solar light emanating from a central armillary sphere, a double reference to Newton's *Opticks* and to his astronomical theories. Reluctantly rejecting that scheme as impracticable, Boullée drafted a second design in which sunlight would shine through small holes pierced in the vault, creating an interior illusion of the domed sky studded with twinkling stars. He intended visitors to enter through the small entrance at the base and then, paralleling a balloonist's sublime flight, experience an ecstatic voyage into celestial infinity.[52]

Boullée and his associates deliberately contrasted the vast arena of life and enlightenment above ground with the subterranean sublime darkness of death. Their designs resonated with Masonic symbolism, and the small, concealed entrances to Boullée's cenotaphs provided limited, privileged access to universal space and scientific knowledge. Boullée was intimately involved in the yearly Academy of Architecture competitions, and even after his death, prizes were still being awarded for Newtonian cenotaphs based on his models.

Preoccupied by death and its representation, several of these visionary architects also chose to commemorate intellectual conquerors of the realm of knowledge rather than military heroes or aristocratic rulers. Just like their poetic contemporaries, they developed astronomical imagery to articulate sentiments of infinity, mortality and human insignificance. For instance, one of Boullée's students was Claude-Nicolas Ledoux, whose dramatic plan for a new cemetery showed the earth surrounded by swirling clouds and planets. Ledoux is now notorious for his extreme version of this type of speaking architecture, a phallic brothel. However, he was also responsible for many of Paris's important new buildings. In his books and lectures, Ledoux deliberately made the word 'attractive' resonate with multiple meanings of

Newtonian physics, chemical affinity, and the visual appeal of beautiful structures.[53]

Newtonian attraction, astronomy and equality became ideologically fused in some Revolutionary imagery. In an early broadside showing the *Astronomical system of the French Revolution*, Equality is one of the forces radiating out like the sun to protect the nation from aristocratic despots and anarchic brigands hovering round the perimeter. The image's planetary orbits of democracy and constitutional monarchy are shown as circular. Circles and spheres traditionally represented perfection, but in Revolutionary France they also became symbols of equality. The designer of a new monument explained in a neat little rhyming epigram that 'a globe is the most perfect emblem of equality because it is always only equal to itself'.[54]

Architects wanted to transform the public spaces and buildings of French cities into physical declarations of circular equality. Every point on a circle's perimeter is equally distant from its centre, symbolizing how in an egalitarian political system, all people are governed by the same laws. In the cross-section plan of his spherical Newtonian cenotaph, Boullée emphasized the circular divisions of the interior space, which dwarfs the Newtonian altar at the centre. Newton's egalitarian gravitational attraction is embodied in the building. His interior design resembled other projects, particularly Jean-Jacques Lequeu's Temple Dedicated to Equality, in which the figure of Justice stood on a globe so that her head was exactly central.

When Louis XVI's scaffold was erected in the centre of the Place des Victoires, its location reversed conventional power symbolism. The guillotine dropped in the place normally reserved for statues of rulers, but now authority had switched to the spectators round the edge. Boullée's colleagues stressed the equality of space rather than the sovereignty of the central figure, so that the middles of their symmetrical squares and circles became ambiguous places. The old astronomical metaphor of authoritarian rule from the centre was replaced by an idealistic vision of individuals subject to equal forces and moving freely through undifferentiated spaces. As a later architect commented disparagingly, 'even in the palaces of kings, you can see a whole crowd thrown together confusedly according to the principle of holy philosophical equality'. Boullée designed a vast and empty cenotaph, whose luminescence modelled the Newtonian light of reason that pervaded a spacious universe democratically governed by mutually interactive forces. 'Temples of death should', wrote Boullée, 'cast ice into our hearts . . . I wanted to place Newton in the abode of immortality, in the heavens.'[55]

6

GENIUS

Genius has privileges of its own; it selects an orbit for itself; and be this never so eccentric, if it is indeed a celestial orbit, we mere star-gazers must at last compose ourselves; must cease to cavil at it, and begin to observe it, and calculate its laws.

Thomas Carlyle, 'Jean Paul Friedrich Richter' (1827)

The last volume of Joseph Addison's *Spectator* appeared in December 1714. Using a typically Enlightenment metaphor of illumination, the dissenting lecturer Henry Grove exclaimed in wonder: 'How doth such a Genius as Sir *Isaac* Newton, from amidst the Darkness that involves human Understanding, break forth, and appear like one of another Species!'[1] Remarks like this, appearing in a widely read journal, helped to establish Newton's reputation among contemporaries who knew little about his theories. Yet although we instinctively empathize with Grove's awe, we would be unlikely to couch our feelings in this way. This is partly because Grove was using a language that immediately strikes us as old-fashioned. But at a more subtle level, we do not even fully appreciate what he was saying. Interpreting such apparently transparent statements involves trying to enter the very different conceptual framework that governed eighteenth-century thought.

Grove and his contemporaries often used words that appear familiar, but have changed their meanings. Genius is now the label for a singular individual who is far removed from the central norm of society. But in Newton's lifetime, it referred far more often to a specific characteristic that belonged to a person, nation or place. The man of genius was blessed by God with an exceptionally large amount of a particular ability, and so was quantitatively rather than qualitatively set apart from his peers. Writers referring to Newton's genius commonly had in mind something closer to what we would call a special talent or gift, which could be for practical as well as mental tasks. Newton was celebrated for having an unusually strong genius for mathematics, but other people also possessed a genius that enabled them to excel in other fields – Pope in poetry, Kneller in painting, or a woman in embroidery.

Grove's understanding of 'another Species' was also very different from ours. He was living in the early eighteenth century, an era when most English people believed that the world had been created by God in 4004 BC, and since then had remained essentially unchanged. Strongly influenced by classical beliefs, Grove and his contemporaries still held the notion that a single continuous chain of being had been established by God at the Creation. Rooted in inanimate minerals, this chain extended through plants and the simplest organisms upwards to more complex animals, with human beings at the summit. Perched on top, a spiritual ladder stretched up towards God, so that newly released human souls could join the angels and other supernatural beings.

Grove envisaged his own posthumous spirit flitting round the universe to attain far greater knowledge than was possible on earth, a style of cosmic voyaging popular among poets. With varying degrees of elegance, Newton often appeared as an eternal traveller whose extraordinary intellectual abilities gave him holy status:

> *By death from frail mortality set free,*
> *A pure intelligence, he wings his way*
> *Through wondrous scenes . . .*
> *Of saints and angels . . .*[2]

When Grove gushed that Newton appeared to come from another species, he inferred that Newton belonged among these semi-divine entities arrayed progressively between people and God.

This vertical positioning is also implicit in Hogarth's *Weighing House*, where the gentlemen are ranked according to how much gravity they have in their heads (*Figure 3.2*). Hogarth – as in so many of his caricatures – has inverted the accepted social order, here by placing the *Stark Fool* with no gravity at the top. Intuitively, of course, one would have expected Newton, whose head was packed with gravity, to occupy the superior place. The sequential upward ordering implied by the chain of being was embodied in many eighteenth-century literary images as well as material designs. Just as God had set human beings above the rest of creation, so too Newton loomed over lesser mortals from his scarcely attainable peak of knowledge:

> *[Newton] whose towering thought,*
> *In her amazing progress unconfin'd,*
> *From truth to truth ascending, gain'd the height*
> *Of science, whither mankind from afar*
> *Gaze up astonish'd.*[3]

In addition to Newton's metaphorical height, this verse also captures the physical feeling of staring up at a statue placed on a tall pedestal, or at a temple of genius mounted on the summit of a hill.

This progressive chain of being has long been replaced by the branching tree of evolution. For us, species are not fixed, but can change into new ones, and (at least in countries with a Judaeo-Christian background) there is generally a firm dividing line between the human race and God. Moreover, we now think statistically, evaluating people's characteristics numerically in relation to the collective behaviour of a social group. Rather than imagining individuals ranged vertically in gradually increasing order, we visualize a population spread out symmetrically and horizontally about a central peak. Intangible mental abilities are, like easily measurable quantities such as height and weight, assumed to be normally distributed. By definition, most people lie in the central band, while those who are more unusual lie further out to the tails on either side. Newton no longer lies near the top of a great chain of being extending from minerals up towards God. Instead, he is now squeezed into the far right-hand tail of a human gamut that ranges from those of low IQ placed at the left, through a diffuse average, on past the high achievers to a tiny group of geniuses.

One way of approaching genius in the eighteenth century is to follow Samuel Johnson's example in his *Dictionary*, and turn to Alexander Pope, Newton's twin Enlightenment hero:

One science only will one genius fit:
So vast is art, so narrow human wit:[4]

Unfortunately, this couplet makes it clear that, compounding the difficulty of understanding older connotations of genius, the words 'science' and 'art' also meant something very different from now. Words that seem familiar can be just as treacherous as the deceptively similar 'false friends' of a foreign language. Pope was harking back to older distinctions between theoretical and applied knowledge, or between book learning and practical skills. 'Science' could cover any field of scholarship, including what we would call the arts, while 'art' often carried slightly derogatory implications of contrivance, unnatural artifice or manual labour.

Considering Pope's verse underlines the fact that we are estranged from his notion of genius not only through linguistic shifts in the word itself but also through profound cultural changes. Traditionally, the accolade of genius had been most commonly reserved for poets, a usage stemming from classical beliefs about divine inspiration. Before Newton joined this heavenly throng, the only two men universally acclaimed as geniuses were Homer and Shakespeare. In the middle of the sixteenth century, the artistic writer Giorgio Vasari had promoted Florence by advertising the city's painters and sculptors. Giving artists greater status served to broaden the group of men endowed with genius, even though they worked with their hands as well as their minds. Another huge shift took place towards the end of the eighteenth century, when scientific practitioners were becoming respected for their expertise, and the whole notion of genius was being reappraised. Newton's followers placed him in a new category: scientific geniuses.

In complete contrast with the situation in Newton's lifetime, genius is now most strongly associated not with literary or artistic creativity, but with scientific originality, particularly in physics and mathematics. This shift is one that Newton – or rather, his posthumous representations – helped to effect. Viewed retrospectively, this may seem like a smooth transition, but in actuality it occurred gradually and unevenly. British people disagreed not only about the rewards brought by Newton's approach to the world, but also about the concept of genius itself. Moreover, the processes of change were not the same in England as abroad. In particular, the story of Newton's celebration as a scientific genius differed in England from in France, the other country that became a leading centre for neo-Newtonian ideas in the early nineteenth century.

Historians are less inclined than philosophers to pose counterfactual 'What if?' questions, but they do sometimes wonder what would have happened if Newton had not lived. Would someone else have introduced the laws of gravity, expounded the spectral nature of light, and redated the chronology of ancient dynasties? Very probably all of Newton's discoveries would have been made sooner or later, but not all by the same person. What is clear, however, is that Newton would not now be celebrated as an undisputed genius if huge social transformations had not occurred since his death.

The turn of the eighteenth and nineteenth centuries was a key period not only for Newton's reputation, but also in the histories of science and of genius. Examining Newton's elevation to the realm of genius is not to question his ability, since he was undoubtedly an extraordinarily gifted man, but to ask why it is that he in particular should enjoy such a special status. That we should celebrate him as a unique scientific genius depends also on the vital importance now attributed to science and technology, and on the almost religious reverence that we attribute to particular types of ability. Far from challenging the value of Newton's innovations, to inquire why we call him a genius deepens our understanding of the role he has come to play in modern society.

Anodyne accounts simply accept Newton as a great – perhaps the greatest – scientific genius who ever lived. But why were Kepler and Galileo not hailed as geniuses in the eighteenth century? Why do women *still* not earn this coveted title? Were inventors like James Watt called geniuses, or was the accolade reserved for scientific innovators? How did poets and musicians reconcile the cult of genius with more mundane concerns about earning a living? By seeking answers to these questions, we can explore why Newton is so important to us.

Classical culture

William Pattison was a Cambridge drop-out who, despite possessing an 'uncommon Genius for *English Poetry*', spent uncomfortable nights trying to sleep on park benches. Despite financial assistance from Pope, Pattison died of smallpox only a few months after Newton, but left behind a glowing tribute to this newly minted intellectual hero among his poems. Doubtless with his own plight in mind, Pattison declared that British people should encourage native talent, and stop relying on the standards of classical civilizations. Newton, he proclaimed, was to be the new national leader:

> *Our Land can furnish Men of Fame,*
> *To eclipse the* Greek, *and* Roman *Name . . .*
> Newton *shall lead our ravish'd Souls . . .*
> *Our Nation's boast, our Country's Love.*[5]

Pattison presciently articulated major transformations that would take place during the course of the eighteenth century. An accomplished classicist, he lived at the peak of what was often called the Augustan Age, when English gentlemen modelled their lives and literature on an idealized view of imperial Rome. Looking back, we might interpret the early eighteenth century as a period of transition, when an older, classically oriented culture was dying out and being supplanted by an industrial, commercial society. But for the participants, no such clear pattern was discernible.

Latin was still the international language of scholars, equally appropriate for scientific, literary or theological works. Augustan schoolboys were routinely expected to round off Latin essays on natural philosophy with a witty poem, and gentlemen – even the very occasional woman – translated important works from English into Latin. Members of learned correspondence circles exchanged Latin verses that mingled discussions of Newton, politics and religion with bawdy humour. Newton himself wrote in Latin, turned to ancient texts for mathematical inspiration, and sat for portraits carrying allegorical references to the past. This allegiance was well known: the posthumously published frontispiece of his book on fluxions stressed how Newton's philosophy was rooted in classical thought (*Figure 3.1*). In 1741, the *Gentleman's Magazine* ran a competition to translate Pope's famous couplet about Newton:

> *Nature and Nature's Laws lay hid in Night.*
> *God said,* Let Newton be! *and All was* Light.

Strange though it might seem to us, they received far fewer entries in French than in Latin (and even one in Greek).[6]

Pattison evidently admired Newton, but during his own lifetime it was Pope, his literary patron, who enjoyed greater fame. When Voltaire visited England, he remarked that 'the portrait of the Prime Minister is over the mantelpiece of his room, but I have seen Mr Pope's in a score of houses'. Although he was often satirized, Pope was also held up as the ideal modern Augustan. His own poetry was rich in classical allusions, and he had become wealthy through translating Homer, frequently admired as the outstanding example of enduring genius.[7] But by the early nineteenth century it was Newton rather than Pope who was regularly included in the roll-call of the country's great men. Nowadays Pope does, of course, occupy a prominent position in the literary canon, but outside academic circles he is far less well known than Newton.

On the eve of the French Revolution, the eminent Edinburgh judge Lord Kames was worried that British mathematics was slipping into terminal decline like the arts of Greece and Rome. Still measuring modern achievements with a classical yardstick, he praised 'the great Newton, who, having surpassed all the ancients, has not left to his countrymen even the faintest chance of rivalling him'. Kames was vastly outnumbered by more optimistic commentators who congratulated themselves on the rapid rate of recent discoveries. Throughout the eighteenth and nineteenth centuries, scientific enthusiasts hailed Newton as an emblem of progress towards a great British future:

> Learning . . . presently runs forwards, with such an amazing Rapidity, that the Modern Improvements therein, have infinitely outdone all the Attainments of the Sons of Men in that Kind before them put together. And this Mighty Progress of the Sciences (with Joy I speak it, for the Sake of my Country's Honour) was Principally owing to the Immortal Genius of the Great NEWTON.[8]

British culture remained saturated with Greek and Roman references. It was only gradually that manipulating equations came to be seen as a more useful skill than declining Latin nouns. Newton was the product of a society that prided itself on its Latinate origins, and he owed part of his fame to being compared with classical heroes whom he was destined to displace. Even when writing in English, Newtonian propagandists constantly included references to people, gods and literature that their readers would instantly have recognized, but which are unfamiliar today.

One particularly significant poem was *On the Nature of Things*, a six-part epic by Lucretius, a Roman contemporary of Cicero and Julius Caesar. Designed to persuade a stressed politician to change his hectic life-style, this long poem expounds the moral and scientific beliefs of Epicurus, the Greek philosopher who had emphasized the importance of searching for personal happiness. Fully translated into English for the first time near the end of the seventeenth century, *On the Nature of Things* provided a model for writers who developed an important genre of English poetry that simultaneously hymned and explained the mysteries of nature. Lucretius's work also appealed to natural philosophers, because their new mechanical, corpuscular theories mirrored Epicurus's belief that matter is composed of tiny separate particles.

Newton became the new Epicurus. Steeped in classical thought, British writers could draw correspondences not only between these two heroes' scientific interpretations, but also between their behaviour. Both of them had proposed atomic theories, and – just as Epicurus had advised – Newton led a secluded life at Cambridge, achieving his greatest insights in the rural tranquillity of Woolsthorpe. It was Halley, Newton's loyal publicity agent, who initiated this process of identification. The long Latin eulogy he composed for the first edition of the *Principia* is laden with Lucretian imagery, and extracts and English translations were frequently printed. The last couplet conveys the flavour of Halley's hymn, which underpins Newton's godlike depiction in *Figure 1.5*:

> *Newton, that reach'd the insuperable line,*
> *The nice barrier 'twixt human and divine.*[9]

By far the most influential passage from *On the Nature of Things* was Lucretius's description of Epicurus's death. In the standard eighteenth-century translation, these famous lines appeared as:

> *Nay,* Epicurus *Race of Life is run,*
> *That Man of* Wit, *who other Men out shon,*
> *As far as meaner Stars the mid-day Sun.*[10]

Here an image of celestial light is combined with death, that favourite topic among Englishmen, who were renowned for their melancholic preoccupation with mortality. This was an ideal image for celebrating the man who had brought order to the cosmos as well as researching into optics. Several poets borrowed and embroidered these lines. For instance, the fifteen pages of verse that prefaced Pemberton's Newtonian primer lavishly embroidered Lucretius's neat metaphor of the Epicurean sun that outshone lesser stars:

Like meteors these
In their dark age bright sons of wisdom shone:
But at thy NEWTON all their laurels fade,
They shrink from all the honours of their names.
So glimm'ring stars contract their feeble rays,
When the swift lustre of AURORA's face
Flows o'er the skies, and wraps the heav'ns in light.[11]

These lines may seem tediously ornate, but the author could be confident that many of his contemporaries would recognize his allusion.

Newton became indelibly linked with Epicurus after 1755, when Lucretius's original description of Epicurus in this quotation – *qui genus humanum ingenio superavit* – was carved on the pedestal beneath Roubiliac's statue in Trinity College antechapel (*Figure 2.9*). This tribute is tricky to translate, partly because of ambiguities in the Latin original, but one modern rendition is 'he who surpassed the human race by his genius'. Presumably selected by the Master, Robert Smith, who had commissioned the statue, this Latin motto reinforced visions of the marble Newton looming over his visitors just as Newton himself towered intellectually over the entire human race. The epigram's Epicurean origins became obscured with frequent repetition, but in illustrated books and articles it came to identify Newton himself as intellectual leader of the human race.[12]

Over the centuries, the relationships between *ingenium* (the nominative of the *ingenio* in the statue's inscription) and its

English equivalents have changed, so that the phrase carries different connotations for us than it did for either Smith or Lucretius. All the translations available in the eighteenth and early nineteenth centuries used the word 'wit', but modern experts usually call Newton's Epicurean *ingenium* his genius. The Latin *genius* and *ingenium* had already started to become twisted together by the early eighteenth century, while two other derivatives of *ingenium*, ingenuity and engineer, were also shifting in significance. The linguistic tangle is hard to unravel, but is worth tackling because it reveals some fundamental transformations in how we esteem achievement. That Lucretius's early translators – who included John Dryden – should have chosen wit rather than genius to describe Epicurus illustrates how they appraised intellectual prowess differently from us.

Genius originally came from the Latin *genius*, a word stemming from the verb to beget, which denoted the small creative or advisory spirit sitting on a man's shoulder. Genius gradually acquired some of the characteristics formerly reserved for *ingenium*, an innate quality or nature. So sometimes genius resembled our word 'character': when Ditton was defending Newton, he pointed out that 'Some People again naturally love to find Fault; their Genius prompts them to snarle and censure.' Not only people, but nations, languages or places could also be described as possessing their own characteristic genius. Thus the English were inherently predisposed by their genius to behave modestly and politely, while the French were excitable and gesticulated widely. Modern garden designers still recall Pope's poetic advice to heed the voice of nature – 'Consult the genius of the place in all.'[13]

Through its links with *ingenium*, genius also came to mean an inborn ability for a specific activity, which could be practical as well as intellectual. Newton could not be celebrated as a scientist in the eighteenth century, since the term had not yet been invented. Mostly, people paid tribute to his mathematical talents. Isaac Barrow, Newton's mentor at Cambridge, told a fellow mathematician that Newton was 'very young . . . but of an extraordinary genius & proficiency in these things'. A decade later, the Cambridge Platonist Henry More commented that 'Mr Newton has a singular Genius to Mathematicks' (although he disparaged Newton's interpretations of the apocalyptic seven Vials and seven Trumpets). One mid-century biographer reported that the young Newton was inspired by observing 'many Persons of Genius engaged in the Business of improving Telescopes'. This example nicely demonstrates important aspects of how genius has changed in significance. Nowadays Newton is himself a genius, rather than the possessor of a special quality; we reserve the accolade of genius for extreme mental rather than manual aptitude – while we might admire the ingenuity of radio telescope technicians, we would scarcely call them geniuses; and the plural 'Persons' indicates that genius in a certain field was a common characteristic, owned to a greater or lesser extent by many people – it was a relative rather than an absolute label.[14]

In 1719, one writer apparently thought that using Newtonian imagery would clarify these complicated etymological changes:

> as Gravitation in a Body (to which this bears great Resemblance) doth not barely imply a determination of its Motion towards a certain Center, but the *Vis* or Force with which it is carried forward; and so the *English* Word *Genius*, answers to the same *Latin* Word, and *Ingenium* together. *Ingenium* is the *Vis ingenita*, the natural Force or Power with which every Being is indued; and this, together with the particular Inclination of the Mind, towards any Business, or Study, or Way of Life, is what we mean by a *Genius*.[15]

Even armed with this supposedly helpful hint, the relationships between words such as 'genius', 'ingenuity' and 'engineer' are hard to disentangle, not least because each of them is constantly acquiring new connotations.

The genius cult

Wordsworth must often have reminisced about his undergraduate days at Cambridge, and around 1840 he added the final two lines to his evocation in *The Prelude* of

> *The Antichapel where the Statue stood*
> *Of Newton with his prism, & silent face,*
> *The marble index of a Mind for ever*
> *Voyaging thro' strange seas of Thought, alone.*

This has become the most famous celebration of Newton as a disembodied scientific genius. It drew on several older poetic tributes to Newton and also reinterpreted Roubiliac's eighteenth-century vision of an eminent natural philosopher (*Figure 2.9*). Whereas Roubiliac had sculpted an Enlightenment gentleman engaged in public discourse and inspired by God, Wordsworth expressed Romantic perceptions of a solitary genius detached from normal life.[16]

Wordsworth and his contemporaries produced new representations of Newton that reflected changed attitudes towards science as well as towards Newton and genius. Although men of science were criticized by Romantic literati, they now played an influential role in English society, and were esteemed for their theoretical knowledge as well as their practical expertise. As science gained in prestige, genius ceased to be restricted to literary creativity, and expanded to include technical inventiveness and theoretical perspicuity. Newton's name had – according to the *Historic Gallery of Portraits* – 'become synonymous with genius'. That this accolade could be paid to a scientific innovator rather than to a literary author reflects the changing status of genius as well as of science. Genius was no longer attributed merely to the possession of superior intellectual capacities, but described singular individuals driven by an inner creative force. Wordsworth himself articulated this shift: 'Genius is the introduction of a new element into the intellectual universe.'[17]

Among the new images of Newton that appeared around the end of the eighteenth century, two in particular illustrate these fundamental transformations. William Blake's striking print, coloured predominantly in shades of blue and green, shows a naked, muscular Newton apparently seated on an underwater rock (*Figure 6.1*). Unlike earlier portraits of Newton, this is not intended as a direct representation. As he arches over into a curve that mirrors the semi-circle on the scroll, the muscular ripples along his back and arms contrast with the angular precision of his bent knee, the triangular dividers and his geometrical drawing. These striking tensions between straight and curved lines, symmetrical and asymmetrical shapes, definite and indefinite forms, encapsulate the confrontation between the powers of reason and imagination. Blake is expressing his dread that imagination and religion will be squeezed out of the world by abstract attempts to circumscribe it mathematically.[18]

Blake's poetry and art are often interpreted as articulating Romantic hostility towards science. Nostalgic anti-technology campaigners have converted him into their champion for a vanished past, often quoting his 'May God us keep / From Single vision & Newton's sleep!' – a snappy sound-bite purloined from a private letter that was not even published. Yet his *Newton*, an enormously complex image open to multiple interpretations, reveals deep ambiguities in his attitudes. Indeed, Newton's face is said to resemble Blake himself.[19]

Blake was himself helping to fashion new connotations of science. Fearing that science with its older sense of knowledge was being constricted, he warned that this shift towards its modern narrower definition signalled a decline in cultural values. These regrets were laced with self-interest, since Blake was concerned to depict himself as a born artist, a genius whose innate imagination placed him above those who relied on rational thought. Angrily annotating his copy of Joshua Reynolds's lectures on art, Blake protested that 'The Man who says that the Genius is not Born, but Taught – Is a Knave.'[20] Like Wordsworth's disembodied genius, Blake's Newton is no ordinary mortal, but a superhuman creature with almost terrifying abilities.

In complete contrast, George Romney's purportedly realistic reconstruction of Newton's prism experiment illustrates other ways in which people were thinking about science and genius (*Figure 6.2*). Romney has painted a moment of discovery, the scientific parallel to a Romantic genius's flash of literary inspiration. As Newton captures the divine light of inspiration, he holds his prism like a magic wand over the small orrery symbolizing his mastery of the planets. Although several artists were challenging the validity of Newtonian optics, Romney carefully reproduced the Newtonian spectrum of colours, and copied the prism from a diagram in the *Opticks*.

The first of a genre that would appear with increasing frequency during the Victorian period, this picture shows Newton as an active experimenter rather than a contemplative scholar, gaining his knowledge from nature rather than from books. As men of science became established, they felt more comfortable about being portrayed engaging in manual activities, formerly deemed inappropriate employment for gentlemen. This picture provides the visual equivalent of the revised versions of history being produced by science propagandists who were creatively converting episodes from the past into momentous discoveries made by heroic ancestors.[21]

Romney intended this picture to accompany one of Milton dictating to his two daughters, and his apparently natural composition is highly artificial. At the time, colleagues admired the simple domesticity of the scene, which anachronistically and improbably shows an elderly Newton at home with his niece and a maid. For some modern viewers, Romney's *Milton* and *Newton* reinforce visions of detached male intellects displaying the truths of nature to appreciative, servile women. To others, Newton appears to be brandishing his prism to ward off the evil influence of buxom, corrupting wenches. Such differences of opinion illustrate how pictures can be constantly reinterpreted. Although we only know Newton through his representations, this flexible ambiguity is precisely what makes the analysis of visual material so valuable for understanding how cultural heroes are formed.

Blake and Romney painted their pictures at a time when fifty years of debates about genius were intensifying to a climax. Picking up on classical imagery, young poets and painters portrayed themselves as exceptionally sensitive, original people fired by an internal imaginative force. This cult of genius stemmed from numerous essays that strongly differentiated genius from ordinary learning or talent. In England and Germany, one manifesto of special genius proved particularly influential – *Conjectures on Original Composition*, published anonymously in 1759 by the elderly Edward Young, internationally celebrated author of the Newtonian *Night Thoughts*. What is genius, asked Young, 'but the Power of accomplishing great things without the means generally reputed necessary to that end? A *Genius* differs from a *good Understanding*, as a Magician from a good *Architect* . . . Hence Genius has ever been supposed to partake of something Divine.'[22]

Wearing exotic clothes and indulging in flamboyant behaviour, self-styled geniuses created not only controversial works of art but also new identities for themselves and their acolytes. Since living the life of a true genius entailed being almost literally consumed by the inner flame of creativity, they were obliged to pay for their unique gifts: tuberculosis and mental instability became almost essential hallmarks of genius.[23] The self-portrait drawn by Blake's intimate friend Henry Fuseli is often interpreted as a key image of genius. Dressed in rough clothes, Fuseli confronts his audience with a saturnine stare in a pose that recalls traditional iconography of God Himself. With this brooding scene, Fuseli encapsulated in chalk Blake's boast that '[h]e who can be bound down is no Genius: Genius cannot be bound.'[24]

Many people strongly disapproved of these new geniuses. Augustan society had been noted for its order and regularity, so that progress was often seen as the product of methodical, careful work rather than intuitive leaps of inspiration. Thus Newton's biographers maintained that he had achieved his great insights not from flashes of inspired creativity but through hard, patient thought. Voicing a common assertion originated by Newton himself, the chemist Joseph Priestley insisted that 'if he had done more than other men, it was owing rather to a habit of *patient thinking*, than to any thing else'. Even among his most ardent admirers, Newton was not necessarily called a genius. Thomas Reid, for instance, founder of Scotland's aptly named common-sense school of philosophy, repeatedly praised Newton, claiming that he was following in the master's footsteps to explore and systematize the human mind. Yet, he warned, 'it is genius, and not the want of it, that adulterates philosophy, and fills it with error and false theory'.[25] In contrast, nurturing individual creativity entailed encouraging diversity and nonconformity. Artists and poets were traditionally held to be infused by divine inspiration, but this verged dangerously near to demonic possession, or being swept away by religious enthusiasm approaching madness. In 1798, a reactionary critic at the Royal Academy exhibition wrote scathingly about 'those enthusiastical geniuses of the age who have yet to learn how to contemplate nature with a correct eye'.[26]

Deliberately exiling themselves from conventional society, geniuses trod a narrow borderline between originality and insanity. The chemist, poet and would-be genius Humphry Davy empathized with the problems of his friend Samuel Taylor Coleridge: 'He has suffered greatly from Excessive Sensibility – the disease of genius . . . With the most exalted genius, enlarged views, sensitive heart & enlightened mind, he will be the victim of want of order, precision and regularity. I cannot think of him without experiencing mingled feelings of admiration, regard & pity.' William Hazlitt was less sympathetic, cruelly attacking the chequered trajectory of Coleridge's career: 'Alas,' he lamented sarcastically, ' "Frailty, thy name is *Genius*!" '[27]

By adapting Hamlet's famous comment on women, Hazlitt emphasized the gender paradox of geniuses who were inevitably male, yet in their behaviour and appearance exhibited the fragile sensibility normally attributed to the female sex. So to call

Newton a genius could imply aligning him with effeminate Romantics, although at the same time genius was an accolade to be reserved for men. Several further ambiguities plagued discussions about Newton and genius. The Oxford Professor of Poetry had spelled out the contrast between academic and literary writing: 'The language of reason is cool, temperate, rather humble than elevated, well arranged and perspicuous . . . The language of the passions is totally different: the conceptions burst out into a turbid stream, expressive in a manner of the internal conflict.'[28] It seemed contradictory to link Newton, the supreme icon of rationality, with poets whose creative thought processes dangerously resembled those of the insane. On the other hand, although Newton should surely be placed above lesser individuals who could only achieve results by diligent application of the rules, crediting him with intuitive inspiration threatened to undermine the rational foundations of science.

Labelling someone a genius is not just paying a tribute to one particular individual, but reflects opinions about the relative standing of different types of achievement. Newton was identified as a scientific genius to distinguish him from other types of people, including women, poets and inventors. Like other social categories, the characteristics of a scientific genius were defined in opposition to those who were excluded. At the same time, Newton was a special case who helped to carve out what it means to be a scientific genius. Many writers on genius used Newton to analyse what the term should denote.

Disappointingly for modern feminists, Mary Wollstonecraft decreed that female geniuses were essentially men. Using an appropriately astronomical metaphor, she exclaimed that Newton 'was probably a being of superior order accidentally caged in a human body . . . the few extraordinary women who have rushed in eccentrical directions out of the orbit prescribed to their sex, were *male* spirits, confined by mistake in female frames'. Entrenched attitudes were hard to change, and physical justifications for women's purportedly inferior mental capacities had a long history. Until the end of the seventeenth century, even women followed Aristotle's pronouncement that their soft, moist, cold brains rendered them constitutionally incapable of serious application. Furthermore, a woman's behaviour was held to be dominated not by her head but by her womb, whose malign influence throughout the body was apparently responsible for all sorts of problems derisively labelled hysteria (from the Greek for womb).[29]

Although women were sometimes said to *possess* genius, this was always in the older sense of a special quality or disposition. A woman might, for example, have a genius for friendship, or for a suitably domestic activity such as painting flowers on to silk. Ironically, it was these apparent compliments that prevented women from entering the restricted circle of intellectual geniuses like Newton. By praising women's special genius for sensitive intuition, polemicists – of both sexes – effectively precluded them from activities that demanded rational modes of thought.

Yet poetry, the traditional genre of creative genius, did seem to offer women the possibility of joining Coleridge, Wordsworth and Blake in the new coterie of geniuses. One woman who tried to breach the barrier was Anna Seward, the 'Swan of Lichfield', whose distinguished circle of friends included Samuel Johnson, Walter Scott and Erasmus Darwin (Charles's grandfather). An erudite critic whose own published poems were highly acclaimed, Seward emphasized how geniuses sometimes seem to erupt from humble backgrounds. The fourteenth-century Florentine painter Giotto is still perhaps the most famous example, but metropolitan gentlemen of this period took great delight in unearthing native geniuses from the British provinces. Seward gave such tales of inborn originality an additional twist. If creativity was inherent rather than acquired, she argued, then women who lacked men's classical education but who expressed themselves spontaneously – like a Romantic genius – could also be good poets. Seward's definition of poetic genius emphasized the traditionally female strengths of 'creative fancy, – intuitive discernment into the subtlest recesses of the human heart'.[30]

The male colleagues whom she valued so highly were

embarrassed by Seward's determination, and they undermined her attempts to be taken seriously as a literary writer. William Hayley composed a poem for her, in which he condescendingly used Newton as a foil to contrast male and female genius. Seward, wrote Hazlitt, drew him to her side 'by the Laws of Attraction', thus implying that she had the occult powers of a witch. Like all women, she would remain a mystery unfathomable by the great Newton himself:

In the wonders of Nature Sir Isaac was vers'd,
But Alas! with the Nine he had little Alliance,
And tho' to the bottom of Comets he pierc'd
He ne'er sounded Woman, that much deeper Science . . .
To contemplate her Genius may charm him who saw
All the Secret Sublime of the Starry Abyss.

Perhaps Hayley's clumsy scansion in the last couplet indicates his emotional resistance to taking seriously her claims that she, too, could be a poetic genius.[31]

Invention and copyright

The London physician and author John Aikin marvelled at the mechanical geniuses he found working in the provinces. During his travels round Manchester, he derided James Brindley, the famous canal engineer, as an illiterate, ill-spoken peasant. He did, however, condone Brindley's habit of retiring to bed for a couple of days when he had a particularly thorny problem to work out. 'This is that true *inspiration*,' wrote Aikin admiringly, 'which poets have almost exclusively arrogated to themselves, but which men of original genius in every walk are actuated by, when from the operation of the mind acting upon itself, without the intrusion of foreign notions, they create and invent.'[32]

Aikin's expression 'original genius' crops up in essay after essay. 'The highest praise of genius is original invention,' declared Samuel Johnson.[33] At first glance, he seems to mean the same as

Aikin, but Johnson's celebrated quip comes from his *Life of Milton*. And that was the very nub of Aikin's complaint – the accolade of original genius was reserved for Milton, Homer and Shakespeare. Aikin had been a prolific medical writer before he turned to more literary genres, which was perhaps why he took the unusual step of insisting that technological innovation paralleled artistic originality.

Many essayists on genius agreed with Young that 'in all the arts, invention has always been regarded as the only criterion of Genius'.[34] Confusingly, like 'science' and 'genius', the word 'invention' was also rapidly changing in the second half of the eighteenth century. For literary philosophers, invention was not about James Watt's steam engines or Richard Arkwright's spinning machines, but concerned plots, ideas and metaphors. Mechanical inventiveness had previously been rooted in its Latin meaning, 'discovery', so that inventors were regarded as finding and revealing God's designs, rather than being praised for their own originality. Far from enjoying the high status they came to command during the Victorian era, innovative practical men were looked down on as mercenary craftsmen or opportunistic projectors. It was well into the nineteenth century before inventors and engineers started to gain some measure of security through the patent system. As late as 1834, Watt's new monument in Westminster Abbey showed him in the traditional pose of a philosopher, seated and wearing academic robes. The inscription commemorated him not as a practical inventor, but as an eminent man of science who exercised his 'original genius' in 'philosophic research'.[35]

In Newton's lifetime, the 'Man of *Business*' had been distinguished from the 'Man of *Genius*' who 'looks down with Contempt on the grovelling Creature whose Soul is confin'd to the same Circle with his Trade'.[36] A hundred years later, the goals of wealth and truth were still divorced. Aspiring professional authors wanted to retain the distinctiveness of genius, since claiming originality bolstered their claims to ownership. Nevertheless, they were also wary of the taint associated with working for money rather than glory. Genius, originality and financial gain sat uneasily together. Just as men of science wanted to distinguish themselves from mercenary inventors, so too, many people felt that a true literary genius should not stoop to request payment. Yet it was precisely that originality of genius that writers were using to claim that they had created their work and should be paid for it. The bitter debates about literary copyright were, like those about patents, concerned with establishing what it meant to be an original genius.

One particularly vocal campaigner was the polemical historian and educational reformer Catherine Macaulay. Beneath Macaulay's statue in a London church, the inscription by a fervent male admirer states, 'once in every age I could wish such a Woman to appear, As a proof that Genius is not confined to Sex'.[37] Like Seward, Macaulay wanted to democratize genius by extending it to cover women and the less privileged classes, and she actively engaged in public debates about genius and originality. Activists such as Macaulay ensured that what might appear to be a rarefied debate about aesthetics was made directly relevant to legal controversies about economic competition. Like other women, she was not disinterestedly fighting for female rights, but was trying to safeguard her income. In company with men who felt that the copyright system offered insufficient protection, she protested that authors were being deprived 'of the honest, the dear-bought reward of their literary labours'.[38]

A caricature called *The Genius of the Times* mocked Romantic geniuses as scheming entrepreneurs unsuccessfully seeking admission to the Temple of Fame at the top of Mount Parnassus (*Figure 6.3*). Resembling the Temple of British Worthies at Stowe, this Temple is surmounted by busts of Milton and Shakespeare, and is mostly populated by great English writers, including Newton. The modern geniuses, on the other hand, opportunistically either rush out of the scene towards the Bank of England, or else struggle vainly up the hillside before tumbling with their books into the waters of Lethe. The central figure in black is the satirical poet Peter Pindar (John Wolcot), who derides the 'pretty troop of Candidates for fame I have left at the bottom of the hill, all for getting Money'. Fashionably dressed, Byron is being hauled along on a barrel, while Walter Scott rides up the hill on his publisher's shoulders, urging him to 'Give all thou cans't, and *let me hope for More* another 2000 for another Lady of the Lake . . .'[39]

Wealthy gentlemen could afford to publish their own literary efforts, however mediocre, but the writers thronging this print needed to earn their living, and they campaigned for greater control over their works. Two particularly famous copyright cases were fought over Thomson's *The Seasons*, the Newtonian poem whose popularity made it valuable. Many commentators contended that by reinterpreting familiar natural phenomena, Thomson had created a unique literary property. By analogy with wealthy proprietors of real estate, writers should be able to reap the gain from their mental harvest.[40]

For the privileged, it was easy to declare that knowledge and innovation should remain uncorrupted by financial concerns. Lord Camden scathingly contrasted inventors and writers with '[t]hose great men, those favoured mortals, those sublime spirits, who share that ray of divinity which we call genius . . . It was not for gain, that Bacon, Newton, Milton, Locke instructed and delighted the world; it would be unworthy such men to traffic with a dirty bookseller for so much as a sheet of a letter press.' Macaulay systematically refuted this privileged view. Far from being inspired by 'the single motive of delighting and instructing mankind', she contradicted, 'NEWTON was gratified with a place and a pension.'[41]

London booksellers wanted to extend the current limited period of protection and acquire perpetual copyright. For literary giants such as Shakespeare and Milton, perpetual copyright would effectively give them a licence to print money, and they finally lost their cause. As writers gained more control over their work, originality – the criterion of genius – also became a guarantee of ownership. From the 1770s to the 1840s, bitter debates about authors' rights ran in tandem with prosecutions over mechanical inventions. Revolving around definitions of genius and creativity, literary plagiarism paralleled industrial piracy, and copyright protection held much in common with the patent system.

As well as money, these controversies were concerned with class differences and the emergence of new professional groups who wanted to distinguish themselves from one another. As these new group identities coalesced, men of science often found themselves lying between literary authors on the one hand and inventors on the other. Aligning themselves with gentlemanly men of letters was socially more prestigious, yet incurred the risk of sullying their self-presentation as precise, dispassionate experimenters. They wanted to borrow the accolades being heaped on inventors for their contributions to the booming economy and Britain's military victories, but at the same time were concerned to distinguish themselves from tradesmen and artisans. The laurels of scientific genius, many felt, should not be awarded to inventors. Summoned as a character witness to defend Watt in a patent case, an eminent Glasgow professor testified that Watt was an elite, disinterested man of science: 'No Man', he asserted, 'could be more distant from the jealous concealment of a Tradesman.'[42]

Newton, Shakespeare and Milton: in the early nineteenth century, these three men were routinely acclaimed as England's greatest thinkers. The economist Thomas Malthus converted these intellectual heroes into a national asset of incalculable worth: 'To estimate the value of Newton's discoveries or the delight communicated by Shakespeare and Milton by the price at which their works have sold, would be but a poor measure of the degree in which they have elevated and enchanted their country.'[43] Although modern commentators would be unlikely to bracket a scientist with two poets, the word 'scientist' had not yet been invented and Newton was often included with men whom we would regard as literary writers. One of William Hazlitt's friends judged that 'the two greatest names in English literature [are] Sir Isaac Newton and Mr Locke'. (As the conversation moved on to Shakespeare and Milton, Newton was discounted on the grounds that he was not a person or a character, but a book – the *Principia*.)[44]

Newton was definitely a national icon, often referred to as a genius, but how did he become a *scientific* genius? Probably the first writer to use the term 'scientific genius' in its modern sense was Alexander Gerard, who belonged to the select 'Wise Club' of scholarly men at Aberdeen. A Scottish preacher who became a professor first of philosophy and then of divinity, Gerard won a prize for his 1759 *Essay on Taste*. He had already started work on his long *Essay on Genius*, but fell back on the familiar excuse of overwork to explain why it was not published until fifteen years later. Unusually for the time, Gerard distinguished two kinds of genius, artistic genius which tries to produce beauty, and scientific genius which searches for truth. Explaining that this contrast is due to different mental combinations of imagination, memory and judgement, Gerard repeatedly cited Newton as his main example of a scientific genius. To capture the distinction between geniuses like Newton and Homer, he used a suitably optical metaphor: 'A penetrating mind emits the rays by which truth is discovered: a bright fancy supplies the colours by which beauty is produced.'[45]

Finding the first instance of a term's use is an enjoyable academic exercise, but does not necessarily yield much information about what happened historically. Intriguing as Gerard's ideas are, it seems unlikely that his book had much effect: it was not reprinted, Gerard himself was rarely referred to, and very little has ever been written about him. More than fifty years after he had praised Newton, the concept of scientific genius still seemed to entail an inherent contradiction: 'Surely, whatever is *invented* cannot be *science*. Invention belongs to *art*, and to the creation of *genius*. Science analyses facts and develops principles. It *discovers*, but it does not *invent*.'[46]

Like many historical transformations, Newton's conversion into a scientific genius took place sporadically and displayed no clear pattern. Words like art and science, genius and invention, imagination and originality, were extraordinarily fluid in the decades around the beginning of the nineteenth century. To borrow a comment that Hazlitt made on originality, the very vagueness of the term genius explained 'how a thing may at the same time be both true and new'.[47] A genius could excel in many different areas and the word genius itself meant many different things. This very multiplicity enabled important shifts in the general understanding of genius to take place, since the term's vagueness allowed writers to slip between one definition and another, even sometimes within a few sentences. As literature and science gradually became carved into separate areas on the map of knowledge, Newton carried the multi-layered label of genius with him into the new domain of science. When science gained in prestige, genius expanded its meaning as Newton, already a great English hero, became a founder member of a new international elite of scientific geniuses. Characteristics like originality that defined literary genius also came to be attributed to Newton, the paragon of objective reason.

One way of illustrating this loose slippage of genius from one context to another is to examine the essays and newspaper articles written by the most influential commentator on genius, William Hazlitt. Hazlitt often drew a sharp cleavage between scientific subjects that demand methodical application, and the fine arts of painting and poetry, which – he claimed – depend on genius. Achieving scientific knowledge is, he explained in 1814, a cumulative process independent of any one individual's mental powers. In contrast, artistic geniuses tower above their peers and endure for ever. Yet despite this concern to separate the arts and the sciences, over the following years Hazlitt's ambiguous phrasing sometimes left it unclear whether he intended to describe Newton as a genius or not. In a famous essay about genius and greatness, he declared that 'lawgivers, philosophers, founders of religion, conquerors and heroes, inventors and great geniuses in arts and sciences, are great men . . . Among ourselves, Shakespeare, Newton, Bacon, Milton, Cromwell, were great men; for they shewed great power by acts and thoughts, which have not yet been consigned to oblivion.' Is Hazlitt describing Newton simply as a great man, or also as a genius of science?[48]

By 1830, Hazlitt judged that genius implies 'the greatest strength and sagacity to discover and dig the ore from the mine of truth'. This clichéd definition seems entirely appropriate for Newton, yet actually he was discussing artists such as Reynolds and Titian. In an essay concluding with a dramatic account of Galileo, Hazlitt wrote that truth is the essence of originality, which is itself the criterion of genius: 'the value of any work of art or science depends chiefly on the quantity of originality contained in it'. This older Hazlitt had evidently become far more willing to juxtapose science and original genius. Perhaps he was thinking of Newton when he wrote: 'The mind resembles a prism, which untwists the various rays of truth, and displays them by different modes and in several parcels.'[49]

Rational progress

The developments of English and French genius were very different, yet in both countries, accolades of Newton fashioned changes in the very concept of genius itself. When the editors of the *Encyclopédie* were drawing near the end of the alphabet, they must have realized that they had forgotten to include an entry specifically discussing Newton himself. Fortunately for them, he had been born in the tiny hamlet of Woolsthorpe, and it was under the heading 'Wolstrope' that a French biographer proclaimed: 'England can be proud of itself for having produced the greatest & the rarest genius that ever existed.'[50] This claim is not quite so straightforward as it might seem. It was designed to persuade readers as well as to inform them, since even in the 1770s some doubtful readers still needed to be convinced that Newton's glory should eclipse that of Descartes.

In France, discussions of Newton's genius were intimately bound up with ideological claims that reason was the motor of progress. As autocratic rulers and churchmen became less powerful, they were supplanted by a new intellectual monarchy led by secular heroes. Newton was elevated to the semi-divine status of a great historical figure, a timeless genius who served symbolic functions.

One of Napoleonic France's most prominent playwrights, Népomucène Lemercier, wrote a poetic epic devoted to humanity's four great men – Homer, Alexander, Moses and Newton. Newton, the rational man of science, now belonged to a collective pantheon of geniuses unlimited by epoch or field of achievement. Like a religious offering, Lemercier dedicated his scientific section to Newton: 'My fable is a homage to your genius'. Titled *Newtonian Theogony*, its rhyming couplets described a fictional Atlantis inhabited by a large cast of mythological characters whose romantic exploits allegorically explained modern concepts. By 1834, at the peak of French Romanticism, Newton and genius had become so automatically linked together that Victor Hugo selected a Newtonian metaphor to evoke a genius's special power: 'through an irresistible law of attraction, all minds gravitated around the radiating mind'.[51]

The French Revolution is often held to be the birth of the modern world, and Newton's political role as the French genius of reason is an important component in this story of his reputations. Just as in England, the word genius carried various implications in France. Its meaning was so contested that the *Encyclopédie* had five separate entries, as well as various definitions scattered within other articles. Quite evidently, contributors held contrasting and often conflicting views of this elusive concept. Many of them were swayed by Diderot, who was probably also the behind-the-scenes author of one of the genius essays. Placing a new emphasis on imagination, Diderot challenged the conventional French view that genius was rooted in rationality. Instead of celebrating geniuses as the favoured beneficiaries of divine inspiration, he suggested they should be revered – and feared – for being driven by their own internal passions. Because Diderot constantly revised his concepts of genius and originality, his arguments were far from consistent, yet he was grappling with precisely the dilemma that lay at the heart of debates about that paradoxical term, scientific genius. How, he wondered, could he reconcile his belief in the power of rational, deductive reasoning with his more intuitive sense that creative thought must depend on making illogical leaps?[52]

In his novella *Rameau's Nephew*, Diderot pushed this interpretation to extremes by depicting a mad genius who dangerously threatened the boundary between enlightened, thoughtful people and the criminally insane. His work strongly influenced Louis-Sébastien Mercier, a prolific writer who is now scarcely remembered, but whose sentimental novels broke sales records in the last third of the century and came to inspire more famous authors such as Hugo. Mercier articulated the central paradox of converting the hyper-rational Newton into a Romantic genius: how can methodical, deductive reasoning be compatible with the intuitive, almost random, juxtaposition of previously unconnected ideas? Mercier presented a Jekyll and Hyde version of Newton. During sleep, his brain is plagued by erroneous ideas and vague phantoms that, like scattered soldiers responding to a drum roll, leap into order the instant he awakes to become, once more, 'the man of genius, who pursued truth with such admirable wisdom'.[53]

But most of Mercier's contemporaries avoided confronting this dilemma. Romantic concepts were taken up far later in France than in either Germany or England, partly because of political differences. It was not until the 1820s that Romanticism, with its emphasis on original creativity, became important, when it was as much a political as a literary rebellion. The young authors who insisted passionately on intellectual freedom of expression were also seeking to overthrow the reactionary power of conservative social institutions. Before then, despite Diderot, writers continued to emphasize that genius was a rational faculty bestowed by God, and possessed by everyone to a greater or lesser extent.

The English and German cults of genius were stimulated by fights to gain copyright ownership for original works, but this impetus was lacking in France, where the state supervised patents and copyrights far more closely. Provided writers and inventors could satisfy the stringent bureaucratic requirements, they benefited from financial assistance and protection. On the other hand, they were also subject to tighter controls. Before the Revolution, the central administration had exerted a tight print censorship by giving the King sole power to grant authors the privilege of publishing their works. Writers could not own or sell ideas, ministers argued, because these were a gift from God: only the King, His representative on earth, could determine what knowledge God had chosen to reveal. Such attitudes obviously dampened enthusiasm for original genius. In 1789, freedom of the press became heralded as an inviolable right, but although the old system was abolished, individual authors still had fewer rights than in England. Truth was now to be found in nature, which was the property of everyone, so that the role of writers was to act as agents of enlightenment by ensuring open access to knowledge. Under this democratic reformulation, originality again became devalued as authors lost their former privileged status to become public servants.[54]

Unlike Young and the other British moral philosophers who stressed originality and creativity, influential French authors maintained that inventive genius was based on skill, ingenuity and – above all – reason. Whether excelling in literary, artistic or scientific spheres, a man of genius was not intrinsically different from his peers, but was able to reason more quickly. By mid-century, Newton had joined an international roll-call of great Enlightenment men inspired by the genius of reason. In one of Mercier's poems, Fénelon and Addison provided convenient rhymes for Newton, as 'Genius spread its divine greatness' over 'a hundred different names nurturing Reason'. This bond was, if anything, tightened during the Revolution, when reason became a virtue to be hymned along with liberty and equality. As the Jacobin Marie-Josèphe Chénier put it, 'It's good sense, reason, that does everything . . . And genius is sublime reason.'[55]

Discussing Newton's genius became important in political and philosophical debates about equality and the relationships between the state, the individual and science. While English monarchs had lent little but their name to the Royal Society, the French crown had always provided financial support for the Paris Academy. This alliance between science and government increased during the Revolutionary period, when politicians and philosophers fervently preached that scientific research promised social improvement. To strengthen their arguments, they made Newton an exemplary member of a new intellectual elite that was to supplant older hierarchies based on aristocratic and clerical power. At the same time, their polemical assertions about Newton's special abilities made key contributions to French concepts of what it meant to be a genius.

Hailing Newton as a unique genius was riddled with ambiguities. How could an egalitarian society fête an exceptional individual? Was Newton born a genius, as some writers maintained, or did he become one? This question later entered Newtonian discussions in Victorian England, and is one aspect of what we call the nature–nurture problem. It became central to French debates about politics and education towards the end of the eighteenth century. If geniuses are born, not made, then progress lies in the hands of individuals rather than the natural evolution of society.

Newton became enmeshed in French radical notions of social progress and perfectibility at two levels. Most obviously, he was the outstanding example of a rational genius in the modern era, the man who had stood on the shoulders of Descartes and other giants to bring order to the cosmos. More fundamentally, many philosophers and men of science believed that Newtonian-type laws could be used to describe the development of living beings and even society itself. Newton himself symbolized the possibilities of progress; at the same time, strengthening his significance, his success at mathematizing the physical universe seemed to guarantee that laws governing social improvement could also be found. Newton the astronomical law-maker provided an icon for human legislators. When the legal reformer Jeremy Bentham had studied French radical philosophers as a young man, he confidently boasted that 'The moral world has . . . had its Bacon; but its Newton is yet to come.'[56]

In his controversial funeral eulogy, Fontenelle had proclaimed that Descartes and Newton 'were geniuses of the first order, born to dominate other minds, and to build empires'. Just as a divinely appointed king was born to rule a nation, so too, God gave a few selected men the gift of genius so that they could conquer the realm of knowledge. Philosophical ideas about genius changed in tandem with political ones about hereditary privilege, as reformers stressed the importance of environment and education. Searching for laws governing social improvement, Condillac argued that climate affects a nation's character and development. Progress, he insisted, depends on accumulating knowledge, so that future generations will inevitably be more advanced. Ironically, Newton was becoming a figurehead who underpinned exactly those progressive models of human development that themselves implied his own replacement in a succession of great geniuses.[57]

Condillac influenced France's leading reformers, particularly the powerful politician Turgot, who was uniquely placed to implement as well as to generate new ideas about science and education. In his blueprint for change, individual geniuses featured as crucial agents of progress, so that intellectual leaders supplanted hereditary rulers. Rejecting older beliefs that some individuals were singled out at birth, Turgot taught that nature distributes ability equally: 'Genius is spread amongst mankind like gold in a mine. The more ore you take out, the more metal you will get.' Because of their social circumstances, he argued, too many people had been unable to develop their natural talents and had died in obscurity. By reforming the state education system, France's gifted youth could be effectively nurtured so that as the population expanded, more geniuses like Newton would be produced. Displacing hereditary rulers, this new intellectual aristocracy would ensure France's rapid scientific and technological progress.[58]

The major manifesto of perfectibility, *Sketch for a Historical*

Picture of the Progress of the Human Mind, was written by Turgot's aristocratic disciple, the Marquis de Condorcet, as he hid in a friend's house to escape the Terror. Published in 1795 shortly after his death in prison, 3,000 copies were immediately distributed as official government doctrine. Condorcet outlined a nine-stage model of human history, in which 'we pass by imperceptible gradations from the brute to the savage and from the savage to Euler and Newton'; at last, Newton appeared to provide 'a law that has hitherto remained unique, like the glory of the man who revealed it'. Concerned with how new inventions affected society, Condorcet looked forward to his tenth and future epoch, which promised unlimited advances in knowledge and virtue. For him, Newton's major contribution to 'the progress of the human mind' lay in his demonstration that precise mathematical methods held the key to discovery.[59]

Condorcet tried to negotiate the apparent contradiction between egalitarian ideologies and the singularity of great geniuses. Mirroring Condillac, he stressed that scientific advance was not inevitable, but depended on a great thinker being in the right place at the right time. 'Had [Newton] appeared earlier,' Condillac had maintained, 'he might have been a great man for the age he lived in, but he would not have been the admiration of ours.' Newton's success, Turgot declared, 'will be useful to us in showing how the happy conjunctions of chance combined with the efforts of genius to lead to a great discovery, and how less favourable conjunctions might have retarded them or reserved them for other hands'. Even an ordinary schoolboy, he rather unrealistically claimed, now knew more mathematics than Newton. Under Condorcet's influence, educational reformers pinned their hopes on fostering incipient geniuses – potential French Newtons – who would, through promoting France's intellectual and technological supremacy, bring equal benefits to all the nation's citizens.[60]

But around the turn of the century, some disillusioned idealists sceptically dismissed Condorcet and the other philosophical reformers of their parents' generation as naive optimists. For this heterogeneous group of social utopians, Newton provided an emblematic father figure who symbolized exactly that Enlightenment emphasis on rationality against which they were reacting. Emphasizing their own poverty and rejection, these mystically inclined young men depicted citizens not as identical atoms democratically governed by the same laws, but as diverse individuals who were spiritually motivated to live together in mutually beneficial communities. Often rather vague and incoherent, their texts were not so much agents of change as symptomatic blueprints of contemporary social attitudes. Two of these visionary idealists – Henri Saint-Simon and Charles Fourier – illustrate the powerful shadow cast by Newton's iconic stature during the confused era between the Revolution and nineteenth-century positivism. Initially worshipping Newton as their intellectual ancestor, they later denounced their hero to present themselves as the new saviours of humanity.[61]

An aristocratic *sans-culotte* who abandoned his titles, Saint-Simon emerged from a short spell in prison to embark on an intensive self-education course in mathematics and physics as an unofficial student in Paris. Evidently with his own experiences of persecution painfully in mind, in 1803 he published his plans to reform society by establishing a secular faith based on Newton's tomb in England, 'that country which has constantly been a refuge for men of genius'. Following God's instructions that he received in a dream, Saint-Simon proposed forming an international organization to provide financial support enabling geniuses (sub-text: himself) to dedicate themselves to their inventions. Unlike Newton, whom he condemned for selling out to become Master of the Mint, these geniuses would work for the benefit of all mankind. Written in Geneva, his hierarchical scheme was strategically – yet unsuccessfully – designed to gain Napoleon's approval through flattery.[62]

In retrospect, Saint-Simon's project seems an almost comical blend of rational egalitarianism and mystical idealism, but it reflects how this post-Revolution generation was preoccupied with finding new routes to social progress. Saint-Simon used

Newton's singular status to inspire an intellectual democracy headed by a small group of geniuses willing to transmute their self-interest into harmonious love. Each financial contributor would vote annually to choose twenty-one great geniuses, who would themselves elect a president. Divided into four international groups, the organization's members would make a yearly pilgrimage to one of the local temples built in Newton's honour. Saint-Simon's imaginary buildings sound very similar to the cenotaphs to Newton designed by Boullée and his followers (*Figure 5.3*): perhaps he had encountered architectural students discussing them during the years he spent in Paris fraternizing with members of the École Polytechnique. Recalling the Masonic characteristics of these architects, in Saint-Simon's temples dedicated to the Newtonian faith a special door provided entry for a privileged intellectual elite. Descending to an underground sanctum, they would participate in the rituals of Newtonian adoration.

Although Saint-Simon never pursued this early plan, he did continue his life-long search for a principle of 'universal gravity . . . the sole cause of all physical and moral phenomena'. In contrast with the emphasis other writers placed on the stability and order that Newton had brought to the universe, Saint-Simon perceived a world riven by conflict and tension. For him, the laws of nature mirrored fundamental oppositions between the governed and the governors, the poor and the rich, the peasantry and the landowners. As his writing became increasingly erratic, he strongly attacked Newton and his philosophy but formulated a dialectical model of progress, in which geniuses such as Descartes and Newton alternately used different methodologies to effect a continual spiral of scientific advance.[63]

At the same time as Saint-Simon was elaborating what he called his gravitational morality, Fourier was developing an 'Analytic and synthetic calculus of passionate attractions and repulsions' based on the primordial force of 'passionate attraction'. Fourier later claimed that, like Newton beneath the apple tree, inspiration struck when he was eating an apple in a Paris restaurant. But although he often cited Newton as a semi-divine authority, he never explicitly spelled out the precise details of his complicated theory of universal unity.

Fourier's version of a Newtonian aether was a harmonizing fluid that bonded people together into concordant groups. Strongly influenced by Ledoux's Parisian architecture, Fourier rejected the large open spaces symbolizing Newtonian equality. Instead, he designed buildings intended to provide a strongly structured environment for his ideal communities, comprised of people chosen for their complementary profiles of passionate attraction. Fourier prescribed a highly regimented life in an ordered environment run with monastic regularity (somewhat contradicting his advertised ideals of voluntary cooperation). According to him, this collaborative, harmonious regime would result in an explosive flowering of French intellect. Since talent is universally distributed, he wrote (echoing Turgot), there were potentially 37 million mathematical geniuses matching the calibre of Newton.[64]

Fourier regarded Newton's gravitational model as a sub-set of his own grand theory. Just as Fourier and Saint-Simon had initially worshipped a semi-divine Newton, so too Fourier's admirers made him the leader of a utopian cult. In the frontispiece of an 1845 journal, Jesus, flanked by classical philosophers, 'offers his hand to the supreme inventor of modern times, to the man who has discovered the law of the kingdom of God . . . the sublime thinker . . .'[65] This was of course Fourier, not Newton, who stands behind him staring blankly at an astronomical model. Although unflattering, this image of Newton illustrates his dual iconic role in France. He personified rationality, and he originated the law-like approach to the physical world that governed research into biological and social development.

During the 1830s and 1840s, Fourier and Saint-Simon's most influential inheritor was Auguste Comte, the self-proclaimed father of sociology. Like those two earlier utopians, Comte used the Newtonian system of a law-governed physical universe as a model for the progress of science and society. According to his secular Religion of Humanity, the dead attain immortality through their influence on subsequent generations, and he designed a positivist calendar with ritualized days of commemoration. Comte canonized the world's 558 greatest people by grouping them into thirteen months of four weeks each (some days were named after two secular saints to allow for leap years and women).

Comte's calendar was heavily loaded in favour of French contributions to progress. Even though Newton was shown in capital letters, the secular equivalent of a Sunday, the philosophy and science months were called Descartes and Bichat. In England, scientists were almost unanimously hostile to Comte's small sect of adherents, yet despite tiny membership figures hovering around 100, they profoundly affected Victorian thought. Frederic Harrison, the most fervent believer of the English positivist faith, preached his evolutionary vision that the human race was developing towards transcendent unity. He explained that writers were wrong to single out Newton for special attention, since 'the History of Science is a fundamental part of the History of Humanity, and the life of no man, however great, should be treated thus'. But hero worship proved hard to avoid, even when it had travelled back to England via a circuitous French route. Despite his warnings about secular sanctification, Harrison called the Comteans' London meeting place Newton Hall.[66]

7

MYTHS

And sitting in his night-gown in his chamber – some say, often with but one leg in his inexpressibles! – even till afternoon – calculating, unable to resist the sudden mathematical thought that would dart through his mind as soon as he awoke – who would not pity the fair one that might have happened to become his wife?

Thomas Cooper, *Oration at the City of London Mechanics' Institute* (1850)

Most people know very little about Newton's physics, but they do know that he watched an apple fall from a tree. Newton, quipped an advertising agent seemingly unembarrassed by painful puns, is the 'British physicist linked forever in the schoolboy mind with an apple that fell and bore fruit throughout physics'.[1] Newton's apple has become an international emblem of scientific achievement, featured in ephemera as diverse as Gary Larson's cartoons, postage stamps, home tapestry kits, and publicity material for Japanese engineering. In the courtyard of India's new Institute of Astronomy at Pune, a stone Newton gazes at an apple on the ground as though it had fallen from the banyan tree above him. With the help of a protective shield and a strong helium light, a cutting from the Woolsthorpe apple tree has managed to survive the climate. When it produced two apples, the tree became a pilgrimage centre for tourists from Bombay, over 300 kilometres away.[2]

Shortly after the First World War, the poet Alfred Noyes published *The Torch-Bearers*, a quasi-religious epic designed to restore England's spiritual values. Noyes aimed to show that the natural truth and beauty of the universe are revealed by science, 'a unity of purpose and endeavour – the single torch passing from hand to hand through the centuries'. William Whewell, one of Newton's most dedicated propagandists, had earlier made this Platonic vision of scientific progress a favourite model for self-congratulatory scientists. In his three volumes of verse, Noyes romanticized this celebratory version of history by portraying scientific heroes – Copernicus, Galileo and Newton – as semi-divine Promethean torch-bearers. Painting the past as a linear success story implicitly suggests that science is preordained for a future of continuous advance. This focus on milestones of achievement converts methodical, collaborative research into a gripping adventure tale.

Noyes envisaged Newton watching the moon 'burn like a huge gold apple in the boughs' above him, and he explicitly paid tribute to the symbolic power of mythical tales:

> *Or did he see as those old tales declare*
> *(Those fairy-tales that gather form and fire*
> *Till, in one jewel, they pack the whole bright world)*
> *A ripe fruit fall from some immortal tree*
> *Of knowledge, while he wondered at what height*
> *Would this earth-magnet lose its darkling power?*[3]

We shall never know for certain whether Newton was inspired by watching a falling apple, but this story's factual truth is less important than its symbolic importance. Myths operate on several levels. The world's oldest myths concerned mysterious gods who lived in another world, yet embodied role models designed to reinforce the hidden codes of behaviour bonding human societies together. For his own allegorical moral tale, Noyes chose the genre of epic, which had first been introduced in classical times. Rejecting traditional depictions of fabulous creatures, Greek and Roman poets portrayed male heroes with prodigious capacities and virtues, thus providing superhumans who could be more feasibly emulated. Biographies of real yet unusual individuals also contained anecdotes that acquired mythical significance. In third-century Greece, Diogenes Laertius's *Lives, Teachings and Sayings of Classical Philosophers* combined biography, doctrine and gossip, a format that strongly influenced Renaissance scholars. For the early Christians, legendary accounts of saints' lives illustrated how their virtuous characters enabled them to relive the life of Christ. Noyes depicted Newton and his other scientific heroes as secular saints to revive a war-shattered society.[4]

Although Newton had himself originated the apple story, it was scarcely known until the early nineteenth century. Isaac Disraeli (father of the future Prime Minister) sarcastically added a decorative touch, asserting that the apple had fallen directly on to Newton's head and knocked his phrenological organ of causality. In the *Comic History of England*, the 1840s equivalent of *1066 and All That*, the *Punch* cartoonist John Leech gently mocked the alleged moment of inspiration (*Figure 7.1*). By then, although cartoons still commented pungently on scientific controversies, they had lost their earlier savage edge (*Figure 6.3*).[5] Newton was never the victim of cruel satires, and this was one of his first public exposures as a figure of fun.

As was first said about Einstein, the appearance of cartoons signifies that a genius has become a legend. Caricaturists rely on exaggerating features by which their subject is already instantly recognizable, but they also contribute to fashioning a famous figure's public image. Newton's apple has become an iconic attribute like St Catherine's wheel or St Jerome's lion. Paralleling religious imagery, the apple serves both to identify Newton and to conjure up well-known parables of his achievements.[6]

But such symbols constantly change and acquire new meanings. During the eighteenth century, Newton's most important attribute was the comet of 1680, but its significance was later eclipsed by the return of the 1682 comet named after Halley. Similarly, Leech's cartoon refers to two Newtonian features that were familiar to Victorian readers but are now largely forgotten – his dog and his pipe. In our own time, the falling apple carries great allegorical power, yet whether or not the story is true seems relatively unimportant. In contrast, when this anecdote first started to gain prominence in the early nineteenth century, it was certainly not regarded as trivial. On the contrary, protagonists hotly contested its veracity.

Newton's apple holds much in common with other romanticized episodes, such as Archimedes' shout of 'Eureka' from his bath, James Watt's childhood fascination with a boiling kettle, or Galileo's observations of a swinging lamp in Pisa cathedral. Like them, Newton's apple has come to symbolize momentous moments of scientific discovery, ones achieved through a genius's flash of intuition. As Noyes understood, such quasi-historical details convert famous heroes into mythological characters who seem to live out society's grand narrative themes. By structuring our perceptions of history and of reality, myths help to create, define and bond communities, most typically by giving them an origin. The falling apple has acquired mythical validity as the founding moment of Newtonian mathematical astronomy.

In his lines on Newton's apple, Noyes posed an important question. Was Newton, as the falling apple myth suggests, instantaneously and divinely – almost magically – inspired by plucking the 'ripe fruit' of knowledge to reformulate the universe? Or – the less exciting version – were his theories the product of long and serious contemplation, as he lay 'dreaming in his orchard . . . wonder[ing] why should moons not fall like fruit'? This problem lay at the centre of nineteenth-century debates about Newton and science. The fierceness of these discussions indicates that far deeper truths were at stake than whether the episode had actually occurred. The apple's implications were central not only to academic controversies about the conduct of science and the nature of discovery, but also to broader questions about class, education and morality.[7]

Scientific knowledge does not simply diffuse outwards from a core elite to reach a wider public. When confronted with challenges to their own opinions, readers respond in different ways, which cumulatively and gradually alter the role that science plays and the types of question it asks. Scientific practitioners are influenced by cultural perceptions of science and its heroes, so that successful books, newspaper articles and – more recently – television programmes affect how science is itself conducted. Some scientists even claim that the media pull of Stephen Hawking, Newton's modern equivalent of a national scientific hero, prevents them from being able to publish papers whose findings disagree with his.[8]

Although we take for granted distinctions such as professional/amateur or specialized/popular, during the nineteenth century these were still being defined. Numerous anecdotal accounts of Newton appeared, which reached ever-widening circles as the availability of cheap books and periodicals expanded. Authors, readers and texts interacted to transform public perceptions, and mythical tales about Newton's life played a key role in shaping the ideological values of Britain's burgeoning scientific disciplines. Different versions were frequently published in mass-produced publications as well as in biographies aimed at more specialized audiences. Because they were both designed and received in different ways, they yield valuable information about how Newton became a national hero as well as an icon of scientific genius.[9]

Reinterpretations of Newton's life were not just ornamental flourishes but were laden with ideological import. Far from being merely amusing tales, anecdotes like the falling apple helped to determine what it means to be a scientist. Inconsistencies and absences cloud factual accounts of Newton's own existence, but his life's legendary reinterpretations had a very real impact on the subsequent course of science.

Trees of knowledge

At the end of the nineteenth century, a fictional 'young lady of colour lately deceased at the age of 14' described how 'We were lying the other day on our back in our orchard, staring up into a streaked-apple tree, and thinking, as the apples fell, of little Isaac Newton and the curious upcomes after falls. From Newton's apple tree our thoughts fell off to Darwin . . .'[10] As society changes, so too do mythical associations. A modern writer would be unlikely to imagine such a train of thought in the mind of an American teenager, while Newton's colleagues would have found it incomprehensible.

Newton originated this anecdote a few months before his death, as he sat reminiscing over a cup of tea in his Kensington garden with his younger friend William Stukeley. Although he could not have anticipated its future importance, he did later tell the same story to at least three other people. This repetition suggests both that he regarded it as significant, and that this man who was allegedly unaffected by public opinion – itself a common myth of genius – continued to be actively engaged in fashioning his posthumous image.

According to Stukeley,

> the notion of gravitation . . . was occasion'd by the fall of an apple, as he [Newton] sat in a contemplative mood. Why should that apple always descend perpendicularly to the ground, thought he to himself. Why should it not go sideways or upwards, but constantly to the earth's centre? Assuredly, the reason is, that the earth draws it . . . there is a power, like that we here call gravity, which extends itself thro' the universe.[11]

Stukeley was particularly impressed that Newton should draw a parallel between an apple and the moon, thus linking an everyday event on earth with the motion of the planets through space. Because we have been brought up in a Newtonian universe, it is hard for us to recapture the significance of this imaginative leap. Many of Newton's contemporaries still clung to Greek models of the universe, which sharply distinguished between the terraqueous globe and the celestial spheres carrying the stars and planets. By drawing an analogy between the falling apple and the circling moon in order to formulate a single law of gravitational attraction, Newton united the terrestrial and the celestial domains and mathematically welded the entire universe into a new structure.

Assiduous research has identified the tree in Newton's Woolsthorpe garden as a Flower or Pride of Kent, now very rare but then a popular pear-shaped cooking apple.[12] Since Newton had carefully studied the details of cider production, he was probably aware of this. However, the apple's specific variety and the tree's precise location are historically less interesting than the anecdote's allegorical associations.

Three hundred years ago, an apple's significance was very different from what it is today. Well versed in classical lore, Newton and his educated friends would all have known that in the first beauty contest it was Venus who won Paris's golden apple. Branching trees had long provided a common image for the growth of human wisdom, and were often used by encyclopaedists to depict their classification systems. A falling apple immediately recalled the Fall of Man in the Garden of Eden, when the serpent persuaded Eve to tempt Adam with a fruit from the forbidden Tree of the Knowledge of Good and Evil. Celebrating the completion of the Westminster Abbey monument, a poem in the *Gentleman's Magazine* described how Newton

> *. . . like* Enoch, *stood,*
> *And thro' the Paths of Knowledge, walked with God.*

Evocatively combining the Garden of Eden with the landscaped avenues of Georgian estates, this anonymous author compared Newton with Enoch, who lived for 365 years, one for each day of the solar year.[13]

John Milton was one of the first writers who mentioned Adam succumbing specifically to an apple, a detail that may have developed because of confusion between two meanings of the Latin word *malum* – apple tree and evil. Enormously influential throughout the eighteenth century, Milton's *Paradise Lost* was first published in 1667, the same year that Newton was secluded in his own earthly garden at Woolsthorpe, retreating like Epicurus from the demands of plague-ravaged Cambridge. Milton and Newton may also have been thinking of Hercules' arrival in the garden of the Hesperides, when he laid down his globe to sit beneath a tree bearing golden apples, then slew the serpent twined round its trunk. Artists portrayed Hercules bowed under the weight of learning, accompanied by astronomers carrying dividers and an armillary sphere, two traditional symbols later associated with Newton himself.

Far from being confined to Sunday sermons, biblical accounts carried the force of reality. The Fall in the Garden of Eden permeated discussions about moral responsibility, progress and truth. Francis Bacon, James I's Lord Chancellor whose ideas effectively became a manifesto for the Royal Society, confronted the problem of making scientific research compatible with faith in this post-lapsarian era. Vividly evoking Adam's actions and the serpent's venom, Bacon exonerated natural philosophers from the taint of corruption. He carefully distinguished their disinterested search for natural knowledge from sinful people's presumptuous lust after moral knowledge.[14]

In religious iconography, the infant's apple indicates that Christ, the Second Adam, will redeem fallen humanity. For Bacon's followers, Newton became a new Adam who would uncover God's mathematical laws of nature. Thomson presented him as the saviour who would explain the cosmos to 'erring Man', the fallen human race, and such scriptural metaphors were still widely prevalent in Regency England.[15] In *Don Juan*, Byron explicitly drew the link between the Adamic and the Newtonian apples:

> *When Newton saw an apple fall, he found*
> *In that slight startle from his contemplation . . .*
> *A mode of proving that the earth turned round*
> *In a most natural whirl called 'gravitation';*
> *And this is the sole mortal who could grapple,*
> *Since Adam, with a fall or with an apple.*

The sardonic humour of the last couplet disguises the severity with which Byron attacked the scientific dream of restoring the lost Garden of Eden. He continued:

Man fell with apples and with apples rose,
If this be true . . .
For ever since immortal man hath glowed
With all kinds of mechanics, and full soon
Steam engines will conduct him to the moon.

Steam travel across the Atlantic was just beginning to seem a realistic proposition, so that Byron's ironic comment on utopian inventions give a topical edge to his doubts about extending physical mechanics to develop a moral mechanics, a set of laws governing human behaviour. Since then, of course, many scientific innovations – notably the atomic bomb – have prompted similar questions about science's scope and power.[16]

During the seventeenth century, scholarship was perceived as an indoor activity pursued by melancholic recluses working in darkened studies: Kneller's first portrait of Newton is a typical example (*Figure 2.1*). By choosing an exterior, rustic site of inspiration, Newton endorsed learning from nature rather than from books. Unusually, his older colleague the Cambridge Neoplatonist Henry More chose to be painted in a bucolic setting, leaning against a tree whose trunk is entwined with ivy. Also born near Grantham, More was much admired by Newton, who located his own greatest moment of theoretical insight beneath a tree in a country garden. In his Neoplatonic interpretations, Newton held what we would call a holistic or ecological view of the universe, in which nature played an active, participatory role. Just as he forged close links between the comets he described in the *Principia* and the earth's vegetative fertility, so too the falling apple could metaphorically represent nature's intervention in the creative processes leading to his gravitational theories. When Stukeley drew Newton clasped in the arms of a multi-breasted woman, he portrayed a child of nature rather than the offspring of learning (*Figure 2.5*). Recalling allegorical pictures of a woman holding a medallion of her dead husband or son, Stukeley's image reinforces Bacon's gendered vision of male natural philosophers revealing the hidden truths of female nature.[17]

Trees resonated with their own national and mystical meanings. Formerly regarded as wild and frightening, by the early eighteenth century forests had become romantic retreats bearing an almost religious significance. Throughout Newton's lifetime, active replanting and conservation programmes were being undertaken for aesthetic as well as economic motives. Individual trees carried special importance. Births and deaths were often commemorated by planting a tree, while particular trees, such as the oak where Elizabeth I learned of her accession to the throne, provided living markers of the nation's historical heroes. Later in the century, as enclosures and industrialization contributed to the decline of village life, English urban intellectuals expressed growing nostalgia for rural idyllic retreats. Britain's most famous mythical sage was Ossian, a Gaelic warrior and bard, frequently pictured beneath a solitary tree. The fact that he was a freshly minted literary hoax only reinforces this symbolic significance, and Ossian was reinterpreted by Romantic writers throughout Europe as an original genius.[18]

Apple trees carried particular connotations of their own, standing for holiness as well as Englishness. Said to have survived the Flood and provided the wood for Christ's cross, they were often planted in monastery gardens. In his *Dictionary*, Johnson quoted John Dryden's lines emphasizing their sturdy independence:

> *Thus apple trees, whose trunks are strong to bear*
> *Their spreading boughs, exert themselves in air,*
> *Want no supply, but stand secure alone,*
> *Not trusting foreign forces, but their own . . .*

Apple trees and English chauvinism went together. A 1646 frontispiece shows clergymen tending an apple tree in the shape of a crucifix, a living emblem of true English religion that is being defended by Charles I against the foreign intruders hacking down branches representing national virtues, such as faith and obedience.[19] At the end of the seventeenth century, it became increasingly patriotic to drink cider and apple-based spirits rather than imported French wine and brandy. Encouraged particularly by John Evelyn, the Royal Society actively promoted projects to plant apple orchards, and Newton corresponded knowledgeably with his colleagues about details of cider production, orchard profitability and techniques for transplanting and grafting.[20]

So for Newton himself, as well as for his contemporaries and successors in the eighteenth century, this apparently transparent tale of innocent discovery concealed multiple resonances that have now faded from collective consciousness. The anecdote of the falling apple not only hinted at key philosophical and theological debates about knowledge and progress, but also evoked a quintessentially English scene of rural contemplation beneath a tree laden with national symbolism.

Tales of genius

'The Philosopher's Faux-pas' may not be great poetry, but its first verse neatly associates Newton with some stereotypical characteristics of geniuses:

> *'Great men,' they say, 'have slender wits,'*
> *At least, they're subject to great fits*
> *Of absentness of mind;*

Published in 1824 in one of the earliest mass-circulation cheap weeklies, this poem was probably seen by several hundred thousand people. By relating a humorous story about Newton's absent-mindedness, it allied him with great minds of the past as well as reinforcing the bonds between genius and eccentricity. It seems to be the first printed version of Newton's mythical pipe, which became so well known that Leech and George Cruikshank both included it in their cartoons (*Figures 7.1* and *7.2*).

Once upon a time, the story ran, Newton abandoned his books to court a beautiful young lady, but was overcome by shyness. Eventually, he raised her hand 'with befitting gravity' (presumably an intentional pun) as though he were about to kiss it – but here the fairy tale drew to a sad conclusion:

> *'Oh, the timidity of some!'*
> *The lady thought, as he begun*
> *To puff away with double vigour; –*
> *When lo! he raised the sweet fore-finger,*
> *(Shame on the man who thus could shock her),*
> *To make it a tobacco-stopper!*

Newton's pipe seems to have even less historical foundation than his apple, unless perhaps it slipped over from his unfortunate second cousin, John Newton, who died 'by a tobacco-pipe breaking in his throat, in the act of smoaking, from a fall in the street, occasioned by ebriety'.[21]

Newton's pipe and apple first achieved fame in the early nineteenth century, when, like the concepts of science and genius, attitudes towards myth and biography were in a state of flux. Newtonian anecdotes held much in common with those being related about other famous figures. For instance, just as Michelangelo's teacher apparently declared that he had been surpassed by his young apprentice, so too Newton's Cambridge professor Isaac Barrow is often said to have resigned his Lucasian Chair in recognition of his pupil's greater ability. Like the apple tree, this version of events was originated by Newton himself, and it seems more likely that the ambitious Barrow was angling for a higher position.

Such stories helped not only to establish Newton as a genius but also to determine what the characteristics of genius are. In Cruikshank's cartoon, the austere intellectual Newton perpetuates the characteristics of the bearded ancient sage whose picture decorates the wall (*Figure 7.2*). While Cruikshank invites us to laugh at Newton's endearing absent-mindedness, he also confirms that Newton is so absorbed in thought that he has no desire to engage in frivolous activities. Frequently retold and reinterpreted, Newtonian anecdotes assumed the qualities of myths that demand – as Coleridge influentially termed it – a 'willing suspension of disbelief'.[22] Like all myths, Newton's famous deeds and sayings – whether real or not – still continue to gain and lose meanings, as they reflect and mould changing social concerns. Although explicitly about events that may or may not have happened, implicitly they express codes of moral behaviour and scientific practice.

Seeking to express eternal truths, Byron, Scott and other Romantic authors gave traditional myths new meanings as well as consciously inventing new ones. Through their interpretations, Faust and Don Juan acquired a special type of reality that differentiated them from other fictional characters. This borderline identity was later shared by Sherlock Holmes, who embodied features of genius frequently attributed to Newton himself. From the opposite direction, anecdotes involving historical people came to form a body of myth that shrouded their actual lives in unreality. These often survived because they served polemical interests. Like Newton's apple, the only thing most people know about Robert the Bruce is that he sat and watched a spider. However, it was only in the 1820s that this hero for Scottish nationalists was brought together with his emblematic spider, which had previously belonged to someone else. During the same period, several scientific myths arose that advertised the overthrow of superstitious beliefs by rational thought – and hence, of course, confirmed the superiority of the educated classes. These included claims that Christopher Columbus had defeated superstitious clergymen and sailors by proving the earth to be round, and the rumours of 'Give us our eleven days' riots, in which ignorant mobs supposedly resisted England's calendar changes of 1752.[23]

Scientific authors were starting to write celebratory biographies that placed great emphasis on momentous discoveries. The very act of converting Newton into a biographical subject consolidated science's prestige, and also boosted his own fame as an English hero. The first full biography of Newton appeared in 1831, when the Scottish physicist David Brewster published a small, accessible hagiography that sold several thousand copies and proved an influential source for subsequent writers. Brewster's *Newton* formed part of a campaign to undermine radical political activity by distributing cheap non-fiction among working families. Along with over fifty other titles, it was marketed in an ambitious and not wholly successful commercial venture aimed at reaching new audiences with drastically reduced prices. Brewster's *Newton* succeeded better among the middle than the working classes, but it was widely read. Further editions appeared in Britain and America, including a revamped version of the remaindered original, and even those reluctant – or too poor – to buy the book itself could read substantial extracts in the mass circulation weeklies that were being founded.[24]

In 1855, Brewster produced a second, far longer and more technical account, which remained the definitive version of Newton's life until the second half of the twentieth century. By then, to make scientific knowledge appear objective and impersonal, it had become fashionable to separate lives from achievements. Like other academics who claimed that only the minds of scientists were interesting, Brewster excluded or even denied some of the anecdotes that he had included in his first biography.[25]

But in contrast with this elite focus on brains, writers targeting other audiences preferred to explore their subject's character and to imbue their narratives with moral prescriptions. Under different interpretations, Michael Faraday – the blacksmith's son who became England's leading electromagnetic scientist – was glorified either as a Romantic genius or as a working-class, self-taught hero. Florence Nightingale, on the other hand, became a role model for girls, held up as a paragon of hard work and self-sacrifice who confirmed expectations of female domestic virtues.[26] Despite the disapproval of Brewster and his peers, versions of Newtonian tales were frequently included in journal articles, books and poems designed for a wide range of readers. Acquiring through repetition the force of mythical truth, they played a vital role in fashioning Newton's image as a scientific genius.

Writers wrapped their moral messages in attractive packages by embellishing their own favourite stories to present different versions of Newton. Because Newton was already a national hero, these narratives consolidated prescriptions for appropriate English behaviour. In addition, by reformulating traditional accounts of creative geniuses, they helped to create a new stereotype, science's equivalent of Shakespeare, Beethoven or Michelangelo. Although these authors interpreted Newtonian anecdotes in different ways, they mostly relied on the same few eighteenth-century publications that had preserved the personal reminiscences of Newton and his contemporaries. Most significantly, Fontenelle's funeral elogy included intimate details furnished by Conduitt, and – in contrast with later encyclopaedias – historical dictionaries had no qualms about mixing character assessments with achievements.[27]

One particular quotation was repeated so often that it became the verbal counterpart to the falling apple. Borrowing from Milton's *Paradise Regained*, Newton had said, 'I don't know what I may seem to the world, but, as to myself, I seem to have been only like a boy playing on the sea shore, and diverting myself in now and then finding a smoother pebble or a prettier shell than ordinary, whilst the great ocean of truth lay all undiscovered before me.'[28] This image became a leitmotif in scientific thought. In 1850, the astronomer John Herschel followed in Newton's footsteps by becoming Master of the Mint. However, he declined to apply for Newton's Lucasian Chair at Cambridge, modestly demurring that he was better suited 'to loiter on the shores of the ocean of science and pick up such shells and pebbles as take my fancy for the pleasure of arranging them and seeing them look pretty'. Byron was less self-deprecating. In *Don Juan*, he first commented about Newton

That he himself felt only 'like a youth
Picking up shells by the great ocean – Truth.'

and then envisaged himself, infused with poetic inspiration, skimming the 'ocean of eternity'. By the middle of the century, this vision of a childishly innocent Newton had become so legendary that John Ruskin asked Dante Gabriel Rossetti to paint the scene for the new Oxford University Museum (perhaps wisely, Rossetti declined the invitation).[29]

Further evidence of Newton's humility was provided by another frequently repeated maxim that 'if he had done the World any Service, it was owing to Industry and patient Thought, that he kept the Subject under Consideration constantly before him, and waited till the first Dawning opened gradually into a full and clear Light'. Another way of admiring Newton's modesty was to cite his famous disclaimer, 'If I have seen further it is by standing on ye shoulders of giants.' This aphorism originated in the twelfth century, and since then had appeared in various guises, including the windows of Chartres Cathedral. Many of Newton's contemporaries would have immediately understood that his reference had biblical overtones. Now a Newtonian cliché, it reassuringly depicts a humble Newton of less than godlike proportions, and also gratifyingly implies that knowledge is cumulative and science is progressing.[30]

This picture of Newton as a methodical, unassuming, patient worker was propagated enormously widely in the nineteenth century. It was reinforced by his dog Diamond, who became, one sceptic noted, as famous as Alexander's Bucephalus. Diamond reportedly upset a lighted candle in the elderly Newton's London house, destroying a pile of papers containing years of work. Contemplating this disaster, Newton stoically replied, 'Oh Diamond! Diamond! thou little knowest the mischief done.' With its resonances of Christ's words of forgiveness on the cross, this catch-phrase became so well known that when Haydon's baby damaged a painting that he was working on, he boasted, 'I behaved so calmly & well that in future I shall rank the affair by the side of Newton's "Diamond, Diamond." '[31]

The accidental destruction of a masterpiece was a favourite myth of solitary Romantic genius. Most famously, Coleridge's opium-induced dream evaporated from his lonely farmhouse when an unexpected 'person on business from Porlock' knocked at the door: the sordid external world had extinguished the creative act. Fire was a particularly potent symbol of lost knowledge. Just as untamed fire could destroy the library at Alexandria or ravage Newton's candle-lit tranquillity, so too the internal flame of genius disrupted normal human behaviour.

Diamond apparently first came to life in the extended footnote of a 1780 poetic travelogue extolling the pleasures of rural England, but the speed with which this tale entered the Newtonian mythology being generated outside elite scientific circles suggests that it was circulating orally well before then. Significantly, Diamond gained prominence during the intense controversy aroused by French claims that Newton had temporarily become insane after losing manuscripts in a fire at Cambridge while he was attending chapel. England's eminent men of science desperately denied these foreign allegations, but rumours ran that although Newton 'did not get into a passion with Diamond, the injury the dog did was supposed to have deranged him'.[32]

Imputations of madness might add eccentric glamour to the reputation of an artistic or musical genius, but were hardly suitable for a genius held to epitomize rationality. To vindicate their denials of Newton's insanity, learned authors assiduously compared conflicting sources of evidence. Brewster flatly denied Diamond's existence, but in the hands of less partial narrators, the fire that had supposedly occurred in Newton's Cambridge rooms became conflated with the Diamond fire in London. Moving and redating the fire made it easier to gloss over Newton's alleged insanity. Nevertheless, Diamond gradually became consolidated into a spaniel called Tray with a black patch under his tail. As parents and teachers emphasized Newton's calm reaction to underline how their young wards should emulate his patient self-control, Diamond featured regularly in children's books and later appeared on stage in George Bernard Shaw's gentle comedy about Newton.[33]

Accounts of famous artists and writers placed great emphasis on childhood, searching for decisive events and premonitory signs of subsequent achievements. By relating similar anecdotes about Newton, his biographers aligned him with great geniuses of the past. Viewed in retrospect, Newton's birth signalled an outstanding future. He was born on Christmas Day, unexpectedly surviving despite his abnormally small size. Newton's own account became legendary – 'he was so little they could put him into a quart pot & so weakly that he was forced to have a bolster all round his neck to keep it on his shoulders'.[34] Reports of Newton's childhood abilities abounded. For example, in a great storm on the day of Oliver Cromwell's death – a suitably portentous date – he reputedly compared the distance he could jump with and against the wind to measure its strength. The sundial, windmill and other mechanical devices that he built as a boy were renowned, repeatedly cited as evidence of his precocious ingenuity (*Figure 7.3*).

A theme that constantly recurred in artistic biographies was the boy (never a girl) from a farming background whose enormous talents are accidentally revealed. During the nineteenth century, one particularly famous example (provided by Vasari) became the subject of several paintings: Cimabué allegedly discovered Giotto drawing pictures on a stone when he should have been tending his father's sheep, and immediately rescued him to be trained professionally. The Newtonian parallel relates how this boy with unsuspected gifts continually neglected the family sheep and other duties, until

> his maternal uncle . . . learning that he is at the market, goes after him to Grantham. But Isaac cannot be found, for some time; and when he is discovered, it is in a hayloft, and he is absorbed in the study of Euclid. The good uncle discerns the soul there is within him; persuades his mother to struggle with difficulty, and send him again to school, that he may be prepared for the University . . .

Although the evidence suggests that Newton's family was relatively affluent, so that his mother would hardly have assigned him mundane farming chores, various versions of the Cimabuean uncle became standard Newtonian lore.[35]

As such anecdotes consolidated Newton's association with artistic geniuses, new versions of traditional tales about scholars' mental preoccupation gained the comical edge that came to characterize the absent-minded professor. When Maria Edgeworth was trying to demolish the myth of Irish stupidity by collecting examples of English blunders, she recorded (or perhaps invented) gossip about 'the practical bull of our great mathematician Sir Isaac Newton, who, after he had made a large hole in his study door for his cat to creep through, made a small hole beside it, for the kitten'. As an Oxford undergraduate, Percy Bysshe Shelley regaled his father with an adaptation involving hens and chickens, and by the middle of the nineteenth century this rank foolishness had become part of Newtonian lore.[36]

Other stories of this type evolved from more solid contemporary testimony. Thus Pope's Newton who worked at the Mint but 'could not readily make up a common account' reappeared two centuries years later in Shaw's play as a comical mathematician who could barely count.[37] Newton was said to ride with his arms sticking out of the carriage windows, or to be incapable of dressing himself properly. Such apocryphal tales continued an ancient tradition of describing scholarly eccentricity, perpetuated by stories of Einstein's refusal to wear socks, Beethoven's carefully cultivated dishevelment, or Kant's obsessively regular walks. Nevertheless, this was tricky territory. Making Newton *too* unconventional ran the risk of validating French accusations of insanity, so loyal supporters often fell back on convenient euphemisms such as 'morbid sensitiveness ... nervous irritability of mind under peculiar exciting circumstances'. This remains an awkward topic for hagiographers, many of whom welcomed the comforting arguments provided by independent researchers in the 1970s that Newton's strange behaviour could be attributed to heavy-metal poisoning during his alchemical experiments.[38]

One cynic sneered, 'Three things proved the divinity of Sir Isaac, for he never spent a thought on love, took very little sleep, and as for his dinner, he never cared for it and often never ate it.'[39] Self-denial has always been strongly associated with holy fasts as well as with intellectual brilliance. The Greek philosopher Carneades, for instance, allegedly became so absorbed in his work that his hand had to be guided to his plate to prevent him from starving, while Ludwig Wittgenstein's colleagues reported that he would dine on nothing grander than a boiled egg. Several stories circulated about Newton's indifference to food, but the favourite one (even repeated by Brewster) was originally told by Stukeley. Getting impatient at being kept waiting in Newton's dining room, Stukeley had devoured a chicken prepared for Newton. When Newton eventually appeared and saw the bones left on his plate, he remarked, 'How absent we philosophers are. I really thought that I had not dined.'[40] According to several contemporary accounts, Newton adopted a light vegetarian diet in his old age, although several portraits do suggest a robust appetite. Newton's abstemiousness became so renowned that some temperance propagandists and dietary experts cited him to promote their crusades against drinking alcohol or eating meat. Other hagiographers converted his restraint into the religious abstention of a fasting saint, claiming that the *Principia* 'was composed while the body was sustained by bread and water alone'.[41]

Newton's mystical aura was further enhanced by an insistence on his sexual abstinence, particularly significant in England where – unlike European educational institutions – monastic celibacy was officially imposed on Oxford and Cambridge Fellows not only during Newton's lifetime, but until the late nineteenth century (one argument for retaining this ban on marriage was that it encouraged Fellows to resign and thus ensured the introduction of new blood). When Newton was grappling with the *Principia*, he complained, 'Philosophy is such an impertinently litigious Lady that a man had as good be engaged in Law suits as have to do with her'; Halley soothingly dissuaded him from 'desisting in your pretensions to a Lady, whose favours you have so much reason to boast of'. This apparently jocular interchange concealed the gendered foundations of academic science, in which knowledge is pursued by celibate males unveiling the secrets of female nature. Newton never married, and solemnly gave a death-bed assurance of his chastity, thus permitting Voltaire to assert with confidence that medical evidence confirmed his hero 'had neither passion nor weakness; he never went near any woman'.[42]

In Victorian England, sexual matters were referred to more obliquely, although rumours did circulate of a childhood romance that came to nought. Some psychological theories held that a man possessed a fixed amount of creative energy, which could be diverted into either sexual or mental productivity; on this model, geniuses like Newton excelled because they refrained from sexual activity. By perpetuating older stereotypes of celibate saints and hermits, this image of an asexual Newton who conserved his powers for intellectual affairs contributed to defining the characteristics of a scientific genius. Cruikshank's caricature (*Figure 7.2*) had been drawn to accompany an article about romantic follies, and Newton became the personification of scholarly indifference to female charms:

> *Cold, Henry, cold, and philosophical.*
> *Thou art a very Newton in thy love.*
> *Come light thy pipe, and take thy lady's hand,*
> *And use it for a stopper – as he did!*[43]

Converting Newton into a genius who was above normal desires also paradoxically opened him to criticisms of seeming less than human. With Newton in mind, a Victorian cynic sneered 'that the more completely a man devotes himself to science he becomes the less a social being; the less, therefore, a man, and the more a philosophical instrument'.[44] Historians still prefer to describe Newton as an unsullied genius who, like a saint, struggled successfully against sexual temptation. They draw on previously unpublished manuscript material to corroborate their conviction that Newton never indulged in sexual activity with either men or women, even though it seems clear that Newton was at the very least emotionally embroiled with the Swiss mathematician Fatio de Duillier.

Voltaire's bizarrely confident pronouncement on Newton's virginity has lent verisimilitude to this myth of abstinence. But it is not only Newton himself who has become asexual. Newton now represents the idealized scientist prepared to renounce human love in favour of the search for knowledge. Similarly, Galileo became a scientific martyr who sacrificed his freedom for the truth. Image-making processes operate in two directions. Embroidered anecdotes about Galileo and Newton have squeezed their own personalities into a scientific mould, but reciprocally, they have fashioned our image of how a perfect scientist should behave. One English television documentary designed for schools presented as fact a romantic drama about the youthful Newton making the right choice between optics and his sweetheart.[45] The bumbling Newton portrayed by Leech and Cruikshank (*Figures 7.1* and *7.2*) was an early version of Tintin's friend Professor Calculus, but during the second half of the twentieth century this charming innocence faded and new scientific stereotypes appeared – Dr Strangelove, the vindictive Cold War maniac, and Isaac Newton, the bad-tempered reclusive alchemist.

Apples and ambiguities

Whatever they might have felt in private, Newton's associates often paid him lavish compliments. Conduitt oilily remarked that Newton's 'virtues proved him a Saint & his discoveries might well pass for miracles'. Insisting that he had often seen Newton laugh, the ever loyal Stukeley maintained that Newton 'usd a good many sayings, bordering on joke and wit. In company he behavd very agreably; courteous, affable, he was easily made to smile.' But in contrast to this virtuous, convivial dinner guest there also existed another stereotype, one that flattered Newton in a different way: the dedicated, solitary, almost superhuman worker. 'So intent, so serious upon his studies that he ate very sparingly, nay, ofttimes he has forgot to eat at all,' reported his experimental assistant; 'what his aim might be I was not able to penetrate into, but his pains, his diligence at those set times made me think he aimed at something beyond the reach of human art and industry.'[46]

With a wealth of such conflicting reports to choose from, Newton's biographers fashioned versions of their subject that diverged to the point of incompatibility. The reiteration of Newtonian anecdotes not only contributed to making him one of the country's most famous men, but also helped to define the nature of Englishness and the characteristics appropriate for a scientific genius. Propagandists were torn between conflicting images. Many of them favoured progressive torch-bearer narratives, in which Newton appeared in an astronomical lineage of heroes, along with Galileo and Kepler. Such accounts suggest that scientific knowledge depends on the inspired insights of particularly gifted men. But this vision of Newton as a creative Romantic genius who transgressed normal boundaries of behaviour was hard to reconcile with another Victorian role model – the industrious worker engaged in methodical, collaborative research.[47]

Through glorifying Newton, scientists sought to legitimate their own status and to prescribe codes of conduct that would shape future research. But discussions about Newton's character extended far beyond circles of practising scientists and their didactic texts. Newton was unique, one author wrote, because his 'reputation . . . not only illumined the study of the recluse and the salons of society, but has penetrated even to the nursery and the cottage'.[48] People who knew little about science marvelled at Newton's Eureka moment beneath the apple tree, or lauded his hard work and humility. Modern books and television programmes delight in portraying Newton as an ill-tempered recluse absorbed in esoteric alchemical activities, but this character was unrecognizable for Victorians. Already a national hero, Newton was converted into a Janus-faced icon: the Romantic genius inspired by flashes of intuition coexisted with the paragon who embodied British moral virtues of industriousness, patience and religious faith.

Because of these ambiguities, Newton's portraits were seen in new and different ways. Kneller's first picture, in which he had depicted a typically reclusive melancholic scholar, was dramatically rediscovered and reinterpreted by Victorian scientists (*Figure 2.1*). After Newton's death, it had passed through his collateral descendants to the family of the Earl of Portsmouth, and hung neglected in their mansion at Hurstbourne Park for almost 150 years. Since no engravings were made, this image of Newton disappeared from public view until it went on public display for the first time at the large Art Treasures Exhibition in Manchester in 1857. Samuel Crompton (grandson of the famous inventor) enthusiastically described to the local Literary and Philosophical Society how he had been so struck by this portrait that he had taken its photograph. Armed with this copy, he commissioned an engraving, whose dedication emphasized that the original portrait, completed only two years after the *Principia* was published, had been 'handed down from Newton himself and represent[ed] that great man at the most interesting period of his life'.

At this time, art critics were chauvinistically obsessed with finding authentic likenesses of British heroes to display in the new National Portrait Gallery. Ignoring (or perhaps ignorant of) seventeenth-century artistic conventions, and influenced by phrenological expectations, Crompton hailed Kneller's sombre scholar as an inspired genius at the peak of his creativity, the 'yeoman's son . . . at work in the wells of truth, and wresting from nature secrets hidden from the foundation of the world . . . a very beagle of truth' with 'a brow that could measure the universe'. Since then, viewers have given it yet a further meaning, regarding it as a picture of 'the mad scientist', a new stereotype that was gaining strength in the second half of the nineteenth century. It remains the favourite choice of modern historians who have inherited Romantic visions of Newton as a solitary otherworldly genius who cared little for his health or his appearance as he strove to encapsulate the mysteries of the universe in a simple mathematical law.[49]

Hard evidence of responses to particular pictures is hard to retrieve, but it seems clear that Vanderbank's frontispiece for the *Principia* (*Figure 2.3*) also evoked mixed reactions, and so could be used to promote different role models. Crompton, who so admired youthful Romantic genius, denounced Vanderbank's Newton 'as a grand and venerable ruin covered with the lichens of time'. In contrast, a Victorian antiquarian lovingly described Vanderbank's depiction of Newton's steady gaze, silver hair and crimson lips, while a journalist for a Christian family weekly reinforced a paternal image of a 'portly [Newton], his locks silvery but abundant without any baldness, with eyes sparkling and piercing'.[50]

Anecdotes played a vital role in consolidating Newton's multifarious reputations, and the apple story became particularly significant. Scholars hunting down early appearances in print have unearthed Voltaire's mention of fruits, a passing reference in Latin by an eccentric Cambridge academic, and the scholarly (though English) *Biographia Britannica*. However, the first writer to elaborate the tale for a popular audience seems to have been Benjamin Martin, who, as a travelling lecturer and prolific writer, was effectively a less learned version of Desaguliers. In 1764, Martin described how Newton 'was sitting alone, in a Garden, when some Apples falling from a Tree, led his Thoughts to the Subject of Gravity'.[51]

But it was not until the 1820s and '30s that the story first achieved prominence as part of the chauvinistic arguments between Newton's French and Scottish biographers, Jean-Baptiste Biot and David Brewster. Biot's 1821 account, first translated into English eight years later, incorporated several anecdotes whose truth was denied by Brewster in his own book of 1831. Inflamed by Biot's allegations of Newton's insanity, Brewster also rejected his account of the falling apple, although Biot continued to defend its validity. This chauvinistic rivalry between Brewster and Biot was not just about facts, but also indicated how the past was being used to create a national scientific heritage. Historians had previously distinguished periods primarily in terms of the head of state, although to talk about – for instance – the Age of Elizabeth did not imply that Elizabeth herself typified her English contemporaries. In the first decades of the nineteenth century, this chronological classification was replaced by broader cultural descriptions, which tried to capture what Hazlitt termed 'the spirit of the age'. Now it was intellectual heroes – Newton, Leibniz, Voltaire – who served to label great eras in their countries' past.

Brewster was quite openly hymning Newton in order to embarrass the government into funding scientific research. During the international slanging match about sensitive issues such as Newton's priority over Leibniz and his later decline in scientific productivity, even British reviewers accused Brewster of distorting the evidence in 'his one-eyed zeal to promote the glory of Newton'. In riposte, Biot derisively dubbed Brewster 'le docteur Dryasdust', presumably alluding to the fictional antiquarian clergyman who appears in several novels by Brewster's close friend and fellow Scot, Walter Scott. In *Ivanhoe*, his novel about the Norman conquest, Scott uses Dryasdust as a foil to defend his own literary invasion of England, claiming that by mixing fact with fiction he could create an ancestral England as effectively as Shakespeare. Biot was therefore commenting on Brewster's colonization of southern culture as well as his dismissive attitude towards anecdotal evidence.[52]

Biot distinguished two biographical styles. Following Plutarch's example, French authors attempted to bring their subjects to life, whereas the English tradition, he claimed, was to relate every fact in minute detail. His own Plutarchian account of Newton's tranquil isolation from the raging plague resembles contemporary French funeral eulogies, which were themselves eloquent mini-biographies. Modelled on classical ideals of natural philosophers pursuing pure knowledge in pastoral solitude, they inferred that French men of science were insulated from Revolutionary disorder and the corrupting demands of public life. In contrast, British scientific writers who were promoting ideals of cooperation maintained that even geniuses needed the stimulation of their colleagues' conversation.[53]

Newton himself had suggested another key moment of revelation when he used a prism to split light into its spectral colours. Some earlier representations had shown him with a prism, notably Roubiliac's statue in Trinity College and the painting by Romney (*Figures 2.9* and *6.2*). But during the nineteenth century, there was only one significant picture of Newton's optical experiment, displayed by John Houston at the Royal Academy in 1870 (*Figure 1.3*). This deceptively simplified illustration of a dashing young Newton corroborates Newton's claims that his experiments provided clinching demonstrations of his theoretical ideas. Houston conveys on canvas precisely the aura of authority that Newton sought to gain by insisting that he had performed a definitive, crucial experiment.[54]

It was the apple rather than the prism that became Newton's major attribute. Victorian biographies generally placed far more emphasis on gravity than light. There were several reasons for this. For one thing, optical experimenters in France and England had conclusively shown that, contrary to Newton's own beliefs, light is propagated as waves rather than particles. Newtonian propagandists preferred not to focus on this weak point. More significantly, by the middle of the century, Whewell and other Cambridge scientists had replaced theology, the old queen of the sciences, with mathematical astronomy. Resonating with religious significance, cosmological theories carried deeper significance than optical ones. Moreover, theory was increasingly being given a higher status in scientific research than experiment, and Newton's gravitational work lent itself better to educational campaigns that methodology took priority over facts.[55]

The falling apple conveniently symbolized a whole cluster of beliefs about achievement. As one of Brewster's critics argued, 'Let the memory of this precious fruit be carefully preserved as an illustration of the eloquence which sometimes lies concealed in common facts.'[56] Newton's apple became a vivid iconographical attribute that placed him in a mythical time and place rather than in seventeenth-century Cambridge or eighteenth-century London. For instance, in the Oxford University Museum (now the Pitt Rivers Museum), Newton, dressed in a schoolboy's clothes, humbly gazes down at his apple as though it had fallen from Heaven (*Figure 7.4*). Newton appears here as one of a saintly scientific brotherhood that includes his arch-enemy Leibniz as well as Galileo and Humphry Davy. Stretching back to Socrates, these men of science are united by their fame but timelessly displaced from the context of their discoveries.[57]

The Oxford Museum was deliberately built to resemble a Gothic cathedral. This was a period when Victorians were turning towards medieval art and literature to recapture the moral values they felt were being demolished under industrialization. Similarly, at Cambridge, Ford Madox Brown immortalized Newton as a saint-like figure, one in a series of stained-glass windows celebrating Cambridge scientific achievement in the dining hall at Peterhouse. Wrapped in a brown gown against a bright blue background, this medieval scholar with pre-Raphaelite hair is absorbed in silent and almost holy contemplation of his apple.[58]

But these deceptively unproblematic visualizations of a mythical moment of discovery in a remote past concealed virulent contemporary debates. In a wide range of printed media, two fundamental themes dominated discussions about the falling apple – the immediacy of Newton's insight and the innateness of his genius. Although apparently about a particular episode, these differences of opinion revolved around two long-standing and still unresolved arguments. What is the nature of originality? And, to express the question in modern terms, does heredity or environment play a greater part in creating a genius?

Inspiration or perspiration?

Declaring that there is no substitute for hard work, the American inventor Thomas Edison remarked that genius is 1 per cent inspiration and 99 per cent perspiration. In contrast, as Romantic authors reformulated classical notions of divine inspiration, they came to view original composition as sudden, unanticipated, almost involuntary. By pinning down discovery to a discrete moment, Newton's apple, Galileo's swinging lamp and Watt's kettle aligned scientific insight with literary creativity. These anecdotes conveyed an image of effortlessness, paralleling Shelley's insistence that 'it is an error to assert that the finest passages of poetry are produced by labour and study'.[59] Just as Methodist mythology described the infant John Wesley being plucked like a flaming brand from the fire to found a new religion, so too Newton was struck by an apple to create the science of mathematical physics. To give this secular saint immediate access to physical truth, maternal Nature replaced the hand of God:

> *As in a child's hand mothers put the clue!*
> *She dropped the apple in a Newton's sight,*
> *As 't were a plaything for her grown-up child,*[60]

But although this tale of instantaneous, almost Pauline, conversion reinforced Newton's status, it conflicted with contemporary ideologies of scientific practice. Elevating Newton into a singular solitary genius ran counter to the concept of cooperative experimental research, and denied the myth of cumulative scientific progress. Making Newton into an intellectual hero who scaled the lofty heights of truth by thinking in isolation posed the same dilemmas that arose in the Victorian vogue for mountaineering. Whether real or metaphorical, celebrating the individual who reached the peak contradicted the teamwork ethic.

Propagandists for collaborative science stressed Newton's dependence on previous knowledge, the help he gained from his contemporaries, and the ways in which modern scientists were building on his achievements. According to them, 'those who estimate Newton the most highly, are those who think least of the popular story of the falling apple'.[61] Rejecting the apple anecdote, they preferred to reiterate Newton's quotations about gathering pebbles on the shore and standing on the shoulders of giants. These endorsed his own modest emphasis that he was participating in a collective project. To illustrate Newton's patience and perseverance, they cited his restraint when confronted with the Diamond disaster.

Because it symbolized instantaneous, solitary discovery, Newton's apple came to be enlisted in class conflict. The most ardent supporters of cooperative science belonged to the British Association for the Advancement of Science (BAAS), established in 1831 to consolidate the status of science and enlarge the constituency of its enthusiasts. It was at the fledgling society's 1833 meeting in Cambridge that Whewell, a founding member, coined the word 'scientist' in a bid to legitimate this new social category. Like the first members of the Royal Society, the founders of the BAAS claimed to follow the utopian principles laid out by Bacon in the early seventeenth century. Since Baconian science stressed that theories were to be formulated on the basis of extensive sets of data, it relied on contributions from large numbers of careful researchers working under the guidance of experienced superiors. As men from less privileged, provincial backgrounds were encouraged to participate in a national project of improvement, it seemed that science was no longer to be the exclusive preserve of those who had been educated at Oxford or Cambridge.[62]

Not everyone agreed that democratizing the search for knowledge would yield valuable results. In an early diatribe against industrialization, Thomas Carlyle – who later became England's most influential writer on Victorian ideals of heroism – was scathing about the new research institutions being set up outside the traditional universities. He used the apple tree story to articulate this elitist scepticism:

> No Newton, by silent meditation, now discovers the system of the world from the falling of an apple; but some quite other than Newton stands in his Museum, his Scientific Institution, and behind whole batteries of retorts, digesters, and galvanic piles, imperatively 'interrogates Nature,' – who, however, shows no haste to answer.[63]

But middle-class philanthropists were concerned to advertise British liberalism and the benefits of wider education. An old anecdote about a gardener borrowing his employer's books to study Newton became converted into rumours that a weaver had mastered Newton's physics. In *Mary Barton*, her novel sympathizing with oppressed industrial workers, Elizabeth Gaskell made it sound as if Oldham weavers regularly had a copy of the *Principia* lying open on their looms. At the new 'godless' University of London, the first Professor of Mathematics, Whewell's former student Augustus de Morgan, boasted, '[i]n no other country has the weaver at his loom bent over the *Principia* of Newton . . . This country has differed from all others in the wide diffusion of the disposition to speculate, which disposition has found its place among the ordinary habits of life.'[64]

Bacon and Newton were sometimes set up as competing figureheads. Members of the BAAS who favoured a Baconian style stressed the importance of experimental, technical work performed by teams of researchers. This model of scientific practice had room for neither individual flashes of theoretical intuition nor esoteric mathematical deduction by a solitary thinker. When phrenologists addressed working audiences, they vaunted Bacon's broad intellectual interests, and compared his majestic brow with Newton's sloping forehead, which condemned him to a narrow life of mathematics.[65]

Leading scientists struggled to accommodate this tension between the BAAS's Baconian ideology and the evident advantages of promoting science through Newton. In Scotland, the main centre of Baconian enthusiasm at the beginning of the century, Brewster's biased adulation of Newton drew sharp criticism from reviewers, yet influenced the gradual decline in Bacon's status.[66] Ensconced in his Cambridge stronghold, Whewell struggled to reconcile these two iconic figures, almost seeming to identify himself with Newton by stressing his own meteoric rise at Cambridge after his arrival as a provincial scholarship boy.

Although revering Newton as a Romantic genius who obtained his extraordinary results through leaps of imagination, Whewell was aware that he had achieved his own success through dedicated systematic study. He promoted university-based scholarship and placed Newtonian astronomy at the summit of his scientific hierarchy, but was also deeply committed to promoting collaborative, methodical research. Prevaricating, Whewell replaced the contentious word 'genius' with the more neutral 'sagacity', and described a Newton who did not watch an apple, but did sit contemplating in his garden. Governed by the torch-of-knowledge metaphor, Whewell emphasized heroic contributions to science, but generalized individual characteristics into an idealized, stereotypical discoverer. This enabled him to compensate for Newton's solitariness by placing him within a historical collaborative community of inspired minds who never actually met one another.[67]

Giving Newton immediate access to truth undercut claims that knowledge could be achieved through human endeavour rather than divine dispensation. In a learned article about probability, de Morgan satirized the Eureka view of science: 'The Newton of the world at large sat down under a tree, saw an apple fall, and after an intense reverie, the length of which is not stated, got up, with the theory of gravitation well planned, if not fit to print.' Or, as he later quipped in a grating anagram: 'The notion of gravitation was *not new*; but Newton *went on*.' When Oxford's Professor of Astronomy opened his scholarly history of Newton's physics with the apple story, de Morgan was so incensed that he wrote a special *Penny Encyclopædia* article in protest (the volume of N for Newton had already appeared, so he had to write his refutations under the heading *Principia*). Although he was a professional mathematician, de Morgan was also a keen historical researcher, who argued that scientific advance depended on learning from the mistakes as well as the successes of the past. For him, formulating new theories involved processes more closely resembling methodical book-keeping than poetic composition, and his Newton went through thirteen years of detailed calculations, false steps and dashed expectations before completing his theory of gravitation.[68]

But it was not just scientists who disapproved of the suggestion that, like Shelley's effortless poetry, discovery occurred instantaneously. Newton's flash of inspiration under the apple tree contravened Victorian insistence on the rewards of industry. Authors writing for less learned audiences emphasized Newton's dedication, incorporating him within a pantheon of British heroes who had achieved fame and wealth through personal initiative, self-sacrifice, and – above all – hard work. Designed for children, *The Triumphs of Perseverance and Enterprise* rather frighteningly exhorted its young readers: 'the more you know of [Newton] the more will . . . your desire strengthen to approach him in virtue, wisdom and usefulness . . . And work is duty – thy duty – the duty of all mankind.'[69]

Proponents of self-advancement for the working classes rejected the apple tree story, with its implication that truth was immediately accessible to the idle elite. The radical shoe-maker Thomas Cooper groaned about how often this well-worn myth was wheeled out:

> *Others the thread-bare story oft rehearsed –*
> *When as the godlike sage of Albion's isle*
> *Beheld the apple fall, – at once dispersed*
> *Were Nature's mists, and, without further toil*
> *Of mind . . .* [70]

Rather than reiterating this account of sudden inspiration, many authors portrayed the 'child that gather'd shells – kneeling beside the sea'. This Newton was an assiduous learner who diligently proceeded step by step, displaying a laudable humility and simplicity. Even children's books often failed to mention the apple tree, preferring instead to tell the Diamond anecdote. In the early twentieth century, Newton's forbearance in the face of disaster was still being taught in a Halifax school where the motto 'self-control' was carved above the entrance. Pupils were inculcated each week with an uplifting moral theme. 'We could have no more fitting illustration for our Thought "self-control," than Sir Isaac Newton,' intoned their teachers.[71]

Writers who did include the falling apple cunningly adapted it to reinforce this image of Newton as an industrious worker who was prepared to learn from the ordinary world around him. Under this interpretation, the possibility of benefiting from such a trivial event offered hope even to children from unprivileged

Figure 3.2 – William Hogarth: frontispiece of John Clubbe's *Physiognomy* (1763). Hogarth probably intended John Clubbe himself to be the horizontal man of good sense.

Figure 4.1 – William Hogarth: *Frontis-Piss* (1763). Hogarth designed this image as the punning frontispiece for a book about the use of points in written Hebrew.

Figure 5.1 – Newton in Senegal: illustration from "Drame Raisonnable" by Jean Delisle des Sales (1777). Newton eavesdrops on the conversation between a merman and an oyster that is desperately reasoning for its life.

Figure 5.2 – Maurice Quentin de la Tour: *Mlle Ferrand méditante sur la philosophie de Newton* (1753).

Figure 5.3 – Étienne-Louis Boullée: first design for Newton's cenotaph (1784). The coloured aquarelle original shows the mysterious light flooding out from the central armillary sphere. A sense of scale is conveyed by the cypress trees and the tiny figure of Zoroaster at the internal altar, praying with his head upturned in adoration.

Figure 6.1 – William Blake: *Newton*. Conflicting evidence makes it impossible to decide whether this rare but well-known colour print was produced in 1794–5 or 1804–5.

Figure 6.2 – George Romney: *Newton Making Experiments with the Prism* (1796).

Figure 6.3 – 'The Genius of the Times' (1812). The men floundering in the waters of Lethe include Samuel Taylor Coleridge, William Wordsworth and Robert Southey.

Figure 7.1 – John Leech: *Discovery of the Laws of Gravitation by Isaac Newton* (1848).

Figure 7.2 – George Cruikshank: *Sir Isaac Newton's Courtship* (1838).

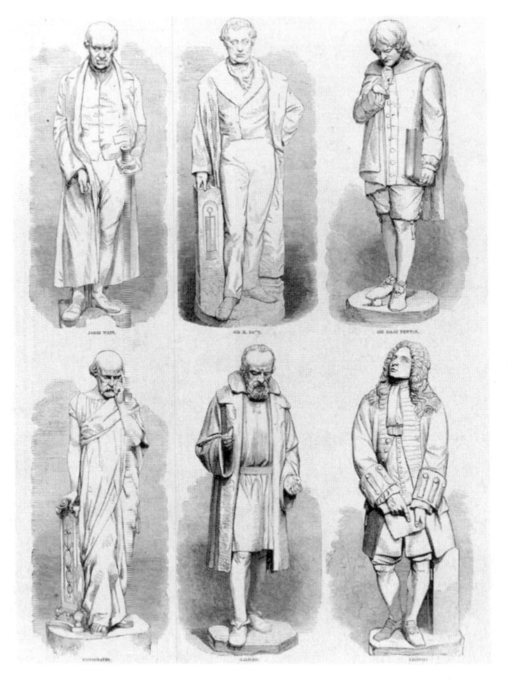

Figure 7.4 – Engraving in the *Illustrated London News* of the first six statues in the Oxford University Museum (1869).

Figure 7.3 – Newton as a child with his windmill and other mechanical devices (1838). Tom Telescope praised Newton's ingenuity by describing his windmill, copied from a real one being built nearby, which was put on top of the house where Newton lodged in Grantham while he went to school.

Figure 7.5 – Mizuno Toshikata:
Isaac Newton (c.1900).

Figure 7.6 – Engraving after Frederick Newenham (1859). *Isaac Newton, at the Age of Twelve.*

Birth-place of Sir Isaac Newton, Woolsthorpe, Lincolnshire.

I have perused yor very ingenious Theory of Vision
in wch (to be free wth you as a friend should be) there
seems to be some things more solid & satisfactory, others
more disputable but yet plausibly suggested & well
deserving ye consideration of ye ingenious. The more
satisfactory I take to be your asserting yt we see
wth both eyes at once, yor speculation about ye use
of ye musculus obliquus inferior, yor assigning every
fibre in ye optick nerve of one eye to have its
correspondent in yt of ye other, both wch make all
things appear to both eyes in one & ye same place
& yor solving hereby ye duplicity of ye object in
distorted eyes & confuting ye childish opinion about ye
splitting ye optick cone.

Is. Newton

Trin. Coll Cambridge
June 20th 1682

Interior of Observatory, St. Martin's St., London.

Figure 8.1 – Newton's three sites of inspiration (1836). Charles Smith published his series of illustrations and facsimile manuscipts as a book of *Historical and literary curiosities*. Other English heroes who he treated in similar style included the artist William Hogarth, the essayist Joseph Addison, and the penal reformer John Howard. Newton's letter about double vision was first published in 1850, and David Brewster later discussed its significance at length.

Figure 8.2 – J. C. Barrow: *Newton's Cottage at Woolsthorpe* (1797). This preparatory drawing for one of Barrow's attractive water colours shows the delapidated state of the cottage at the end of the eighteenth century.

Figure 8.3 – William Theed: bronze statue of Newton. The Grantham Newton is almost thirteen feet high and weighs over two tons.

origins. It could also be converted into sentimental poetic sermons about being content with one's lot. Just as Newton had 'not scorn'd the falling apple', so too should humbler readers trace out 'the golden threads / Of everlasting Beauty' in their 'little dealings, humble trades [and] small besetting cares'. Like de Morgan, these moralizing authors preached that such an everyday occurrence could only be interpreted by a carefully prepared mind that 'had already been devoted for years to the laborious and patient application of the subject of gravitation'. Advocates of the rewards of labour warned that this fortuitous observation had initially led Newton to erroneous conclusions, and emphasized that he had won the prize of truth only after years of intensive research.[72]

Industriousness was tightly allied with religious observance. An article on Newton appeared in the very first issue of a new journal launched in 1832 by the Society for Promoting Christian Knowledge, and three years later his statue at Trinity dominated the front cover. This was a deeply religious Newton, a humble man who gathered pebbles on the shore and dedicated his life to demonstrating God's power and wisdom. Insight only came to the mind that was religiously as well as intellectually prepared:

> *Newton believed for many a year before*
> *The Hand in Heaven shook the Apple down.*[73]

For evangelical believers in free will, individual effort paved the route to salvation. In monthly instalments, many thousands of Methodist children read about 'the Christian Philosopher' and learned that work was almost more important than faith: 'In the path in which Divine Providence had placed him, he [Newton] walked with great carefulness, acknowledging God in all his ways. He was successful, because he was diligent . . .'[74] Newton provided not just an intellectual figurehead but also a moral exemplar, 'a *good man*, as well as a great philosopher', who assumed saintly proportions: 'if all men were to follow the example Newton has set them, it would make a world composed of good and therefore happy men'.[75] Even his death conformed to evangelical ideals, as reports praised his stoic tolerance of increasing pain, his lucidity and his desire to be united with God.[76]

In these didactic reinterpretations, the falling apple symbolized not inspiration but perseverance and humbleness. The mythical potency of this anecdote explains how it could be transported between different communities both within England and abroad. Probably through a translation of Samuel Smiles's *Self-help*, by the end of the nineteenth century Newton had been converted into an appropriate role model for young Japanese people (*Figure 7.5*). This was a period when Japan was turning to Europe for ideas as well as inventions, adapting them to fit the country's own cultural heritage. Pictures of Newton show him beneath delicate un-English apple trees; in one, he is dressed as a smart Victorian gentleman in a checked waistcoat, sitting beneath a Smilesian caption translatable as 'Isaac Newton studies profoundly, but was never boastful'. As well as Newton, Japanese artists painted scenes showing the momentous discoveries of men like Watt and Benjamin Franklin. Although their pictures used different conventions from those in the West, they powerfully illustrate how myths can survive and evolve.[77]

Although its original associations are now concealed beneath layers of accumulated collective memory, the myth of Newton's apple is still acquiring new connotations. In the 1970s, British left-wing activist Dannie Abse, disrupted standard expectations by portraying a mundane hero who suffered from stomach complaints.

Or Newton, leaning in Woolsthorpe against the garden wall,
forgot his indigestion and all such trivialities,
but gaped up at heaven in just surprise, and, with
true gravity, witnessed the vertical apple fall.

O what a marvellous observation! Who would have reckoned
that such a pedestrian miracle could alter history,
that, henceforth, everyone must fall, whatever
their rank, at thirty-two feet per second, per second?[78]

By endowing Newton's laws of gravity with political implications of equality, Abse's radical poem illustrates how anecdote carries the polemical import of myth.

Are geniuses born or made?

As Benjamin Haydon toured northern industrial towns during the early 1840s, he showed his audiences a drawing of a pig's head. Could any of them believe, he inquired rhetorically, 'that gravitation was or could be discovered by a skull and face like these?' At the same time in Newcastle, phrenology students from a far less comfortable background were also being introduced to Newton. What was so special about his early education, they were invited to wonder, that inspired him to discover the laws of the heavens? Nothing, came the answer: like Shakespeare, Milton and the untutored Robbie Burns, Newton was born a genius. People took phrenology very seriously, partly because its political implications conveniently worked both ways. For the privileged classes, distinguished features confirmed a sophisticated brain and an appropriate station near the top of society; for radical artisans, a skull revealing inner native talents implied that a new hierarchy should be built around innate ability rather than inherited status.[79]

Writers across the social spectrum hesitated to emphasize the innateness of Newton's genius because it precluded the possibility of personal development. A few conservative polemicists did find it advantageous to portray Newton as a born genius, because that enabled them to deny the advantages of expanding the education system. Thus the Tory *Quarterly Review* sneered at the '"village" Newtons' being discovered, exclaiming that 'Heaven-born geniuses may take advantage of the opportunities which Mechanics' institutions afford: but they are not dependent on them.' Similarly, Carlyle included Newton (and Faust) in a catalogue of innate geniuses, insisting that science advanced in 'obscure closets' wherever 'Nature . . . had sent a gifted spirit upon the earth.'[80]

But for the majority of Newton's propagandists, to claim that he was born a genius would have been to deny the value of the improving enterprises in which they were involved. Most authors compromised by making the child Newton an unusual boy whose true genius had flowered only in the fertile soil of the English educational system. Such vegetative imagery had pervaded discussion of genius since the mid-eighteenth century, and Newton's apple itself provided a novel if rather mixed metaphor for one popular writer: 'The fall of the apple from a tree in the orchard . . . was the mustard-seed out of which ultimately grew the grand theory of universal gravitation.'[81] In every sort of publication, writer after writer repeated Brewster's pronouncement that Cambridge was 'the real birth-place of Newton's genius'.[82]

Whewell naturally seized on this judgement to reinforce Cambridge's reputation as a unique centre of excellence. Henry Brougham, leading light of the BAAS and the Society for the Diffusion of Useful Knowledge, also insisted that Newton's genius did not show itself at an early age. In this way he could call Newton 'a matchless genius' and yet still encourage apparently less gifted researchers to achieve highly by imitating his industriousness. In the hands of campaigners like Brougham and John Timbs, an influential scientific journalist, Newton demonstrated that committed study ensured progress: in his book about the schooldays of eminent men, Timbs went as far back as the Druids to boast about the efficacy of English education.[83]

In contrast with Newton, men who had succeeded despite their impoverished backgrounds were being converted into mythical heroes for the aspiring classes. Watt, Burns and James Ferguson the ploughboy astronomer (who mapped the stars with a string and beads while guarding the sheep) all provided hopeful evidence that hidden talent could be discovered in the most unpromising circumstances. Some biographers exaggerated the hardship of Newton's rural upbringing, using his rescue by his benevolent Cimabuean uncle to ally him with famous untutored geniuses. Elite counter-propagandists dismissed these 'producers of mere chap-book literature [who] tell droll tales of the herdboy's earliest fortunes': they were delivering a political message of potential success to those born without a silver spoon in their mouth.[84]

In his lectures at the London Mechanics' Institute, the journalist Thomas Hodgskin – who influenced Karl Marx, among others – elaborated an environmental thesis that geniuses are the product of their age. Watt, he taught, should be bracketed with men like Newton as 'master-spirits who gather and concentrate within themselves some great but scattered truths, the consequences of numberless previous discoveries which, fortunately for them, are just dawning on society as they arrive at the age of reflection'. On this egalitarian model, being hailed as a genius meant being in the right place at the right time.[85]

Children's authors struggled to make Newton appear the same yet different from other children. Only by being impressively special as well as comfortingly ordinary could he provide a realistic role model, a paragon whose spirited independence and patient diligence could be emulated by any boy. In these versions, Newton was 'at first very inattentive to his studies, and very low in the school'. His intellectual brilliance only emerged after a legendary playground fight from which he emerged victorious, determined to work hard and achieve his full academic potential. Improving books included numerous examples of Newton's childish ingenuity, especially his sundial, water-clock and mouse-operated windmill. In words and pictures, they showed him actively engaged in devising and making these contrivances (*Figure 7.3*). By stressing his technical bent, writers allied Newton with the inventor heroes of the working classes. In addition, they used him to illustrate the importance of maintaining health by balancing intellectual inquiry with physical activity. They also ingeniously converted this mechanical skill into evidence of Newton's intrinsic moral superiority, since he supposedly made these toys to amuse his classmates while they indulged in idle play.[86]

The longest-running children's book about Newtonian science featured a young lecturer called Tom Telescope. When he first appeared in 1761, Tom was a young, joking boy who stood on the table to teach, and whose precocious cleverness was probably intended to mimic Newton's own. By the final edition of 1838, childhood had come to be revered as a special state of innocence, and successive editors had converted Tom into a more sedate adult. Newton himself was now shown as a robust, sensibly dressed child, surrounded by his own inventions but engaging in the boyish activity of playing with a mouse (*Figure 7.3*).[87]

This small sketch was very different from a large mezzotint engraving that appeared two decades later (*Figure 7.6*). Copied from an oil painting by a fashionable women's portrait artist, it depicted the twelve-year-old Newton meditating with his head on his hand, the traditional pose of reflection. Designed for richer purchasers from a higher social stratum, this picture shows Newton reclining on a tree trunk in his dark velvet suit, his wavy blond locks suggesting submission to the close attention of a mother or servant. His feminized appearance reinforces the innocent purity of childhood and the intuitive grasp attributed to Romantic geniuses, who were inevitably men despite their female characteristics of intuition and imagination. This effete Newton gazes at the stars much as he had contemplated the falling apple – alone and lost in thought.[88]

8
SHRINES

Genius, or, more properly, inspiration, *dignifies every spot on which its energies have been elicited; the dome of the* philosopher *becomes, in record,* academic *and its site may be termed* classic-ground.

European Magazine, 1811

The spark of Romantic genius could be ignited only in a suitably holy environment. A prolific and popular writer on men of genius explained how Buffon had gained his inspiration: 'A secluded and naked apartment, with nothing but a desk, chair, and a single sheet of paper, was for fifty years [his] study; The single ornament was a print of Newton placed before his eyes – nothing broke into the unity of his reveries.' Deep religious significance has often been attached to specific places. For the early Christians, shrines did not merely mark where saints were buried or had performed miraculous deeds, but were themselves sacred groves imbued with holiness. In Newton's England, particular places or countries were characterized by their own genius, a metaphorical legacy from the classical small guardian spirit or *genius loci*. After Newton's death, Stukeley painted the side of his house with a life-size profile portrait of his hero gazing out over the Lincolnshire landscape, 'underneath inscribed GENIO LOCI'.[1]

In March 1927, 150 mathematicians and physicists converged on Grantham, the small Lincolnshire town where Newton had attended school, to commemorate the 200th anniversary of his death. They laid wreaths round his statue, listened to the Bishop of Birmingham preach at a religious service, and made what they called 'a pilgrimage' to Newton's Woolsthorpe cottage, about 7 miles (11 km) away. Einstein sent them a letter, a reverential message that arrived too late – in retrospect, an ironical reminder that even the master of relativity had problems keeping track of the date. 'You have now assembled in Grantham,' he wrote, 'in order to stretch out a hand to transcendent genius across the chasm of time, and to breathe the air of the precincts where he conceived the fundamental notions of mechanics and of physical causality.' Einstein's bid to extend his own hand through space and make his presence felt during the ceremonies exemplifies his self-presentation as the natural inheritor of Newton's 'transcendent genius', a claim to superhuman intellect now being made by Stephen Hawking in an apostolic succession of secular sanctity.[2]

As Newton became England's most celebrated scientific saint, his manuscripts and personal possessions acquired both symbolic and commercial value. Eager collectors pushed up the prices of supposedly authentic Newtonian memorabilia, or gained prestige by donating them to an important institution. An early gift of a lock of Newton's hair to Trinity College prompted further bequests from generous alumni seeking posthumous glory, and by 1815, the going rate for Newton's hair was three times that for Charles I's. The proliferation and circulation of these Newtonian collectibles resembled the treatment of relics of medieval saints, whose holy significance made them valuable commodities to be purchased, stolen or exchanged. Reports that an aristocrat bought one of Newton's teeth for £700 were presumably exaggerated (even though Newton had apparently only lost one tooth by the time he died), but in 1859, a scientist proudly gave the Royal Society an original death-mask he had picked up for a few shillings. Small, cheap versions of Newton's death-mask were being advertised in the press by phrenologists, while custom-made hand-finished models were valued as gifts for scholars throughout Europe.[3]

In the early nineteenth century, British historians with an antiquarian bent became obsessed with tracking down the original manuscripts of famous authors. Eminent men of science quarrelled about their heroic forefathers by retrieving unpublished documents and using them either to blacken or to glorify reputations. Connoisseurs invested famous signatures with talismanic significance, and – compounding the problems of modern researchers – snipped off the bottom of letters to add to their collections. In 1836, an enterprising artist catered for this cult of the past by publishing a series of illustrated tributes to famous Britons, including Newton (*Figure 8.1*).[4]

This composite image illustrates how fashioning Newton's new status as a scientific genius entailed associating scientific innovation with artistic originality. The sheet is dominated by a reproduction of Newton's handwritten letter. Although its contents must have been unintelligible to many readers, the Cambridge source and carefully replicated signature authenticate its value. Above, the reference to Newton's birthplace reminds informed viewers of his near-miraculous origins, a frail infant born on Christmas Day. Woolsthorpe, site of Newton's greatest insights, is shown as a humble provincial refuge that recalls Coleridge dreaming of Kubla Khan in his country farmhouse or Robert Burns scribbling poetry in his remote Ayrshire cottage. Romantic composition in rural solitude provided an artistic blueprint for Newton's portrayal as the solitary thinker who had created a new model of the universe under the tree in his country garden. At the bottom, Newton's bare London observatory resembles contemporary drawings showing Shakespeare's empty room as a hermit's cell. This spartan attic provided an appropriate setting for a genius to retreat from his metropolitan duties at the Mint and engage in nightly communion with the cosmos.[5]

Woolsthorpe, Cambridge and London: these three places of commemoration resonated with Newtonian associations and became secular shrines for intellectual pilgrims. Developers of these rival heritage centres preserved and displayed Newton's saintly relics, and encouraged the ritualizing processes inherent in establishing a scientific faith based on intellectual achievement. But the disparity of these three images reflects how Newton could be promoted as a national genius while simultaneously serving specific local interests. These sites of memory were also sites of conflict that reflected wider debates about the nature of genius and the practice of science. Scientists have actively sought to retain or even invent a continuous relationship with the past through ceremonial practices such as awarding commemorative scholarships, holding anniversary conferences and naming buildings. By appointing their own secular saints associated with places and rituals of worship, scientific communities have aimed to strengthen their internal ties as well as enhance their public profile. Monuments supposedly bond a group's members together by recalling famous heroes or particular episodes communally felt to be important, but they are often erected not simply as reminders, but to instruct people about a particular rendering of history. Protagonists vying to become the guardians of Newton's memory disagreed about which version of Newton should be venerated. As they wrangled, they created different visions of a legendary scientific ancestor from England's distant past.[6]

Cambridge: ancestor of an intellectual clan

When the third annual meeting of the BAAS was held in Cambridge in 1833, participants visited Roubiliac's statue of Newton that had so captivated Wordsworth (*Figure 2.9*). One admirer was Wordsworth's close friend William Sotheby, a distinguished author and Fellow of the Royal Society. Although less gifted, he composed his own hymn to sanctify Newton's genius, marvelling in Trinity's antechapel as . . .

> *. . . o'er thy head*
> *The day's last light, a saint-like glory shed . . .*
> *On thee, prophetic Sage! I awestruck gazed,*
> *Thee, whose pure hand the veil of nature raised*
> *When heaven illumined, thy genius first displayed . . .*[7]

It was at this Cambridge conference that Coleridge goaded Whewell into coining the word 'scientist' to unify what he called 'the empire or commonwealth of science'. Whewell wanted to give men who practised science a collective identity that would distinguish them not only from writers, artists or musicians, but also from those less dedicated and educated people whom he hoped would provide large and appreciative audiences for science. Women, of course, fell into this category. Several devoted wives had accompanied their husbands to the BAAS, but their role was to admire rather than to speak. Although some women were starting to establish successful careers as scientific writers and teachers, they were still excluded from Oxford and Cambridge. As one of these authors self-deprecatingly explained, she might 'have perseverance and intelligence but no genius, that spark from heaven is not granted to the sex'. Whewell and his listeners envisaged scientists as men, and Cambridge itself had nurtured England's greatest scientist of genius – Newton.[8]

Apart from his brief retreat to Woolsthorpe during the plague, Newton spent thirty-five years at Trinity College. Since then, many Cambridge scholars and institutions have claimed affiliation with this famous alumnus. Wealthy benefactors established Newton studentships during the nineteenth century, when the Isaac Newton University Lodge of Freemasons boasted the Prince of Wales among its brothers. Nowadays the Newton Institute for Mathematics displays its maquette of Paolozzi's statue with pride (*Figure 1.2*), although it is hard to judge how much self-conscious irony lies behind the notice-board displays of apple recipes.[9]

But although Newton's fame is distributed throughout Cambridge, it is Trinity College that has most assiduously fostered its historical associations. 'The places where one was young have a great influence on one's life,' pronounced Trinity's Master Amartya Sen, as he posed for photographers in front of Thornhill's portrait of Newton, commissioned almost three centuries earlier by one of Sen's predecessors, Richard Bentley.[10] Using Newton to reinforce its prestige as the University's richest and largest College, Trinity has helped to fashion his public image as a scholarly genius, one of Cambridge's most important intellectual ancestors.

In the mid-eighteenth century, as described in Chapter 2,

Robert Smith, who succeeded Bentley as Master, enrolled Newton to boost Cambridge's standing as an international centre of excellence. Smith wanted to enhance Trinity's reputation as a progressive, successful institution whose own former Fellow had created the new natural philosophy being taught in the modernized educational syllabus. Newton's physical presence is now overshadowed by tributes to other famous Trinity men who succeeded him, but ever-swelling crowds of tourists admire his statue in the chapel, the stained-glass window in the library, and the busts and pictures that Smith either bought or persuaded Trinity men to donate.

Privileged visitors shared the experience of an undergraduate during Smith's regime who was invited to Newton's rooms, where 'every relic of [his] studies and experiments were respectfully preserved to the minutest particular, and pointed out to me by the good old Vice-Master with the most circumstantial precision'. The College was also building up a small collection of Newtonian memorabilia. One of the most prized was an original death-mask, taken by Rysbrack and used by Roubiliac and Romney to create their posthumous images of Newton. However, it seems that some relics only later acquired a special significance: in a 1790 guidebook, locks of Newton's hair and three of his mathematical instruments are listed indiscriminately among intriguing curiosities including a dried human body from Madeira, a giant kidney stone and a papal indulgence. By the 1970s, the authenticity of this hair had become crucial in assessing claims that Newton had suffered from mercury poisoning.[11]

Particularly during Whewell's twenty-five-year reign as Master, in the nineteenth century Trinity College continued to be converted into a shrine for venerating academic ancestors. Most Victorians believed they would be eternally reunited with their loved ones after death, and staunch University men regarded their Colleges as extended families. Whewell explained condescendingly that it might not occur to ordinary people 'how close are the ties which connect a Fellow with his College . . . the tie is closer even than that of the natural family. With those who thus belong to us we share as it were an *intellectual* blood.' As a young undergraduate, Whewell had reconciled himself to spending some of his scarce funds on Newton's *Principia* – 'a book that I should unavoidably have to get sooner or later', he commented to his father. This reluctant purchase proved a valuable investment, since Whewell dedicated much of his academic career to promoting Newtonian methods of science and mathematics. Although mocked by his students as the whimsical Professor F. Uel who experimented with old buttons and ginger-beer caps, Whewell reinforced the College's reputation as an elite hothouse where aspiring scholars could literally follow in the hallowed footsteps of their distinguished forefathers.[12]

As he struggled to define an identity for scientists, Whewell was constantly torn between worshipping Newton as a singular inspired genius who transcended normal codes of behaviour, and extolling him as a man who, though blessed with unusual sagacity, provided an appropriate moral exemplar for less gifted researchers working collaboratively and methodically. Despite these vacillations, it often served his own and the College's interests to portray Newton as a unique product of the Cambridge education system, an outstanding genius hailed as the intellectual patron saint of mathematical astronomy, itself the summit of the scientific hierarchy. In a widely reported public speech, Whewell spoke of Trinity's Newtoniania with reverence: 'We still point out his chambers – we shew his works, his manuscripts, his instruments, locks of his silvery hair.'[13]

Like medieval saintly relics, the College's unique possessions reinforced its threatened authority in Victorian England. As national campaigns to reform the education system gathered strength, staunch College men emphasized the value of traditional methods and sought to refute accusations that Cambridge was old-fashioned and monastic. Defenders of the *status quo* implied that wisdom could be absorbed just by breathing in the University's holy atmosphere, enriched by the spirit of scholarly predecessors: 'the *genius loci* finds an utterance and exerts an agency . . . As we go beneath "Bacon's mansion", or about Milton's mulberry tree; as we kneel where Newton knelt, or dine in halls where the portraits of Erasmus, and Fisher, and Taylor, look down upon us . . .'[14]

Other Trinity Fellows wrote scholarly texts that perpetuated this shared intimacy with Newton in an eternal collegiate fraternity transcending time. One of Whewell's contemporaries scoured College records to publish details of Newton's eating and drinking habits that were inaccessible to less privileged researchers. However, he failed to resolve one problem that continues to perplex Trinity hagiographers: which rooms did Newton occupy at different stages in his College career? In 1963, Lord Adrian – then Master of Trinity – pursued this question in a scholarly inquiry that underlined the proximity of his own study to where Newton might have lived: it is hard to disagree with his endearing conclusion that 'it all sounds too good to be true'.[15] In the 1990s, scientists analysed soil samples in a frustratingly inconclusive attempt to determine the exact location of the Trinity College laboratory in which Newton carried out his alchemical investigations.

Cambridge life is structured by ceremonial practices that re-enact an idealized scholarly past, and Trinity still pays homage to Newton. In 1992, the 350th anniversary of his birth, a commemoration feast was held in the dining hall, which is decorated with icons of Trinity men – including Newton. As the Fellows enjoyed a five-course banquet, in which the Kentish fruit salad was presumably an exotic reference to Newtonian apples, the choir sang *Newtonian Enigmas*, a poetic tribute specially composed for the occasion by the Nigerian novelist Ben Okri. Employing the language of religion, this secular hymn sanctified Newton by imbuing his optical experiments with biblical resonances of the Light of God:

It is not in the nature of Light
That it grows.
But yours does . . .
Amen.[16]

Grantham: local hero or national genius?

The birthplaces of famous people started to become pilgrimage centres during the second half of the eighteenth century, when British tourists became interested in exploring their own country rather than embarking on a European Grand Tour. Many of them flocked to Stratford, where the actor David Garrick was nurturing the nascent Shakespeare industry. More esoteric travellers undertook longer journeys in their search for sources of inspiration, although reality did not always match expectations. John Keats, for instance, eagerly anticipated visiting Burns's remote cottage, but he later complained that 'The Man at the Cottage was a great Bore with his Anecdotes . . . O the flummery of a birth place! Cant! Cant! Cant! It is enough to give the spirit the guts-ache.'[17]

It was a far less successful poet, Thomas Maude, who first advertised Newton's Woolsthorpe cottage in his didactic tours of rural England (*Figure 8.2*). 'When I passed the threshold of his house,' gushed Maude, 'methought I stood on Ether . . . Imagination viewed the Philosopher ranging universal space . . . When I entered the room where his infant eye first saw that Light which he so accurately defined, I was pervaded with enthusiasm.' At the end of the eighteenth century, the word 'enthusiasm' still carried strong religious connotations, and Maude's description evoked the transcendental experience of entering a sacred space.[18]

Although it looked so idyllic in pictures, this remote farmhouse still had its windows boarded up to economize on window tax and bore 'as melancholy and dismal an air as ever I saw'. Even after it was spruced up in the early nineteenth century, few visitors embarked on the three-hour round trip from Grantham. As coachmen ferried travellers along the Great North Road, they pointed out this holy birthsite, extolled by Brewster:

Here Newton dawned, here lofty wisdom woke,
And to a wondering world divinely spoke . . .
All hail the shrine! All hail the natal day![19]

Modern Grantham seems prouder of its second famous native, Margaret Thatcher, whose childhood home has been converted into the Premier restaurant. The daughter of a local grocer, this former chemist boasted about her bust of Michael Faraday, the blacksmith's son who symbolizes rags-to-riches scientific success stories. In her coat-of-arms, she paired a Falklands Admiral with Newton, another Grantham child who made it to the top, thus enlisting him in her politicized manipulation of England's cultural heritage (*Figure 1.1*). Yet in the town itself, the most prominent tribute to Newton is the dilapidated Isaac Newton shopping centre.

Like Thatcher, Victorian propagandists also promoted Newton as both a local hero and a national exemplar for the aspiring classes. By the beginning of the nineteenth century, Newton had become a standard advertisement for England's cultural heritage, used to boost the reputation of home-born talents such as 'HENRY PURCELL, who is as much the pride of an Englishman in Music, as Shakespeare in productions for the stage, Milton in epic poetry, Lock[e] in metaphysics, or Sir Isaac Newton in philosophy and mathematics'. The English nation was being defined not by allegiance to its monarch but by the achievements of its members. As the source of knowledge shifted from faith to reason, some of the emotional commitment formerly dedicated to saints was becoming transferred to the country's national heroes.[20]

Professional associations were also developing – scientists, architects, engineers – whose members were bonded by their work. They often set Newton in a progressive lineage that included Galileo, Kepler and Laplace. On the other hand, English scientists prefer to place him with Darwin and Faraday in a distinctly national scientific brotherhood. One journal candidly confronted this conflict of allegiances: 'Science is of no country; – but nevertheless we cannot help feeling some national pride in the circumstance that physical astronomy was the creation of British genius.' Significantly, this reference to Britain rather than England comes from a Scottish source.[21]

Even as nationalist sentiments strengthened, many people remained loyal to their own towns and regions. Edmond Turnor, the wealthy inheritor of Newton's home, emphasized Newton's Lincolnshire connections and traced his local ancestry back to the early sixteenth century. Turnor's Newton is not only the area's most distinguished son but also an absentee landlord concerned about sheep grazing, whose heirs dispensed £20 to the parish poor.[22] Turnor's brother Charles later erected a prominent obelisk in the grounds of their estate. Now in the car park of the National Union of Teachers, this symbol of eternal fame urges local inhabitants to 'recollect with pride that so great a philosopher drew his first breath in the immediate neighbourhood of this spot'.

Woolsthorpe held double significance, since this place of Newton's birth was also, like the shrines of the early saints, blessed with its own holy tree marking Newton's inspiration as a young man. Shakespeare's mulberry tree generated impossibly enormous quantities of ladles and snuff-boxes, but the Newtonian apple tree industry was conducted at a more refined level. Visiting on a pilgrimage from France, Biot 'gathered a few leaves to carry them religiously back home', while even the sceptical Brewster visited twice and purloined some of the roots. De Morgan maliciously remarked that he 'must have had it on his conscience for 43 years that he may have killed the tree', but whatever Brewster's culpability, by the middle of the century the original tree had blown down and been converted into a chair.[23]

Just as the apple tree locates Newton in mythical rather than factual space and time, so Woolsthorpe has come to typify an idealized English rustic past. Acquired at a low price thanks to the Turnor family's generous and patriotic determination to prevent the cottage from being exported, during the Second World War the Royal Society donated Woolsthorpe to the National Trust. Even though the nineteenth-century extensions had substantially altered its appearance, Newton's cottage symbolized the English values under threat. In 1942, the third centenary of his birth, *Country Life* extolled 'the simple mathematical symmetry of that creeper-covered little manor house . . . this sacred shrine . . . What a grand and typically English countryside this is . . . [that] played its part not only in giving an impetus to the boy's intellectual urge, but also in providing him with that quiet, modest, engaging character that so endears him to us.' Well before Thatcher encouraged the heritage industry to focus on preserving the mansions of famous aristocrats, the National Trust was far more interested in attracting visitors to its splendid country homes and gardens than in commemorating British cultural achievements. It converted this ordinary farmhouse into a romanticized distillation of rural Englishness, decorated with furniture of diverse styles and periods that were mostly unconnected with Newton.[24]

Funding was one of the problems faced by the nineteenth-century publicists who were trying to establish Newton as a Lincolnshire hero. The Lincoln Mechanics' Institute welcomed a donated bust and invited visiting lecturers to talk about Newton, but a scheme to install his statue in Lincoln Cathedral was deemed too expensive.[25] In 1854, Grantham town council started to raise money for a statue commemorating Newton's arrival at the grammar school 200 years earlier (*Figure 8.3*) As with other monuments to men of science, debates about its construction resonated with tensions that spread throughout Victorian society.[26]

State support was one of the central issues. Scientific campaigners complained that the state's unwillingness to erect a memorial to Newton reflected England's embarrassingly poor international record for financing research. Unlike in other European countries, in England public monuments were mainly constructed through private patronage, and the government cited Newton as a precedent when refusing to fund a monument for Faraday. The government did allocate money to commemorate soldiers and statesmen, but middle-class people were becoming more interested in celebrating civic heroes. Capitalizing on this public concern, Brewster stressed the value of scientific rather than military achievement by declaring that 'Newton's glory will throw a lustre over the name of England when time has paled the light reflected from her warriors.'[27]

The drive to commemorate scientists and inventors was boosted by Prince Albert, who recommended founding the National Portrait Gallery to celebrate the country's achievements. Arguments raged about who should be included. Amid rumours of corrupt selection procedures, Newton became an uncontroversial candidate, prominently displayed in preliminary exhibitions. Partisan journalists proclaimed that 'We have outlived the age in which only statesmen and heroes were publicly honoured, in which poets were thrust into the darkest corner of a cathedral, and the professors of science venerated only by pedants and scholars.'[28]

Newton's statue also highlighted tensions between metropolitan and provincial interests. Many British statues were – like Newton's – organized by local groups who combined patriotic sentiments with the desire to embellish their own city. Writers for the national press ridiculed claims that this statue's site was 'consecrated' by the master's footsteps. Derisively bracketing Grantham with other towns trying to profit from their famous citizens, they insisted that 'the partial tribute of a mere local memorial cannot discharge this long-neglected debt of the English people. England cannot do justice to herself except by rearing, in the Metropolis itself, a great and glorious monument.' In contrast, the *Manchester Guardian* defended regional autonomy and provincial civic pride. The city already boasted an expensive, privately financed marble statue of its own famous chemist, John Dalton, and the newspaper maintained that some eminent men would prefer their work to be appreciated by local supporters rather than receive the 'colder but larger admiration' of a London monument.[29]

These conflicts simmered during the four years it took to build the statue. Grantham council raised £600 from local patrons, but almost twice as much poured in from individuals throughout the country. Queen Victoria donated a Russian cannon captured during the Crimean war to be melted down for the statue –

Grantham's Newton was a national hero whose monument literally embodied England's military triumph. In London, the Royal Society refused to contribute, but the BAAS virtually took the project over. The opening ceremonies were timed to follow immediately after the BAAS conference, and the President (the naturalist Richard Owen) played a prominent role. By the time the statue was completed, the unveiling ceremony had turned into a national event. The circulation of engravings, books and newspaper articles helped this provincial memorial to become the property of the entire country.[30]

The statue's appearance must have been hotly debated. Because engravings were relatively cheap, local sponsors of statues were familiar with their hero's appearance. Rather than the idealized classical figures favoured by connoisseurs, they demanded a realistic representation. As phrenology reached its peak, people felt that a sculptor should capture a subject's personality by faithfully depicting his features and cranial characteristics. Newton presented a problem for practising phrenologists, because his death-mask revealed a sharply receding forehead that made his profile closely resemble that of a native American Indian. Unfortunately, this left no room for his organ of causality, but one phrenologist devised an ingenious solution: by comparing Newton's head with that of a blind but stupid woman, he conveniently demonstrated the superiority of Newton's observational powers.[31]

Art critics were often scathing about the style of privately funded memorials. The selection committee picked William Theed, a society sculptor trained in the classical style who provided statues for Osborne, the royal residence on the Isle of Wight. Theed effectively reconciled the subscribers' conflicting demands by working with a death-mask to produce a Newton who held a traditional pose but had a realistic face. Reporters stressed that the Grantham statue was a true likeness, and glowingly described the 'brow grandly developed and beautifully proportioned' that indicated 'the unmistakeable presence of a lofty and commanding intellect'.[32]

Most nineteenth-century scientific statues celebrated men for their practical inventions rather than their theoretical ideas. In contrast, the Grantham Newton displays a planetary diagram and stands near a pile of books that dwarfs his prism and lens (*Figure 8.3*). Swathed in an ambiguous gown, this University-based forefather stems from some indefinite time in the distant past but points the way for his Victorian successors. Theed's statue embodies the Cambridge vision of a magisterial leader for mathematical astronomy, the distinguished ancestor of a collegiate scientific hierarchy.

This figure is strikingly different from the Newton who appeared at Oxford two years later (*Figure 7.4*). Ruskin intended his pantheon of scientific saints to provide students with daily inspiration as they studied, and was scathing about conventional statues that would be ignored by passers-by as they hurried through desolate town squares.[33] Dressed in a schoolboy's clothes and staring down at his apple, the Oxford Newton appears to be a born genius whose flash of inspiration had given him immediate insight into the truths of nature. Paradoxically, the Museum presented students with an image that implicitly minimized the role of systematic scholarship. But the Grantham statue was larger than life, an imposing sage centrally placed so that local people could admire the bronze Newton gazing along the road the human child had taken to school. It was through hard work that this Newton launched his successors on the scientific march towards truth and material progress. He provided an appropriate role model for that local schoolgirl, Margaret Roberts, who successfully upgraded herself to Baroness Thatcher.

The London *Times* snidely remarked that the statue's unveiling 'drew all the notabilities of science together in this unpicturesque, brick-built Lincolnshire town'. The well-orchestrated media event focused attention on Lord Brougham, prominent founder of the BAAS, whose name headed the list of national subscribers. Advance publicity advertised Brougham's visit, front-cover magazine pictures showed him presiding at the ceremonies, and his speech was widely printed in newspapers and in the souvenir pamphlets that sold out the same day. One national headline ignored Newton and ran: 'Lord Brougham in Lincolnshire'.[34]

The statue's inauguration was an extraordinarily colourful occasion. Like triumphal royal entries into Renaissance cities, this secular festival inherited ritual features that traditionally characterized religious ceremonies. It was a bright autumnal day, and a military band accompanied the crowds who thronged the streets of Grantham carrying banners and flags. At the head of the procession marched Brougham, Whewell and Owen, national scientific celebrities who relegated local dignitaries, such as Grantham's Mayor and the Bishop of Lincoln, to second place. Assuming the role of altar boys, three children from Newton's old school reverentially carried cushions bearing treasured relics – Newton's original prism, the reflecting telescope he had made with his own hands, and a copy of the *Principia* – that were placed on altar-like tables in front of the statue. Brougham climbed on to a small dais, which had been brightly decorated in the coloured stripes of Newton's rainbow, to sit in Newton's own chair, a deliberately unrestored 'honoured relic' with the stuffing showing.[35]

Ostensibly this was a joyous ceremony of ritualized bonding, when men of science convened on Grantham to celebrate one of their founding fathers and revel in the achievements of an English hero. But this widely publicized local event gave the participants the opportunity to address the entire nation, and they presented conflicting views of how and where science should be practised. In his speech from the platform, the elderly Brougham undercut Grantham's intellectual claims to Newton by transforming him into an incipient genius who had flowered at Cambridge. Brougham preached his own visionary ideals of democratic education by holding Newton up as a scientific exemplar for the people of an industrious nation. He made Newton conform to his own 'Law of Gradual Progress', which governed moral as well as scientific approaches to attaining the goal of perfection. Placing Newton within a continuous scientific lineage, Brougham asserted that Newton's greatest achievement was to enable 'his successors to occupy what he could only descry'.[36]

Other interest groups also managed to promote their own concerns. Although the Royal Society had refused to give any money, its Newtonian memorabilia had been escorted from London for the opening rituals by the Assistant Secretary, who reported himself delighted to escape from his normal chores for a day. After a '*déjeuner*' for 400 in the Corn Exchange, the future President took the opportunity to improve the Society's public image, currently slightly tarnished. Responding to slurs that scientific status was gained by wealth rather than merit, he promised that he would head an independent organization, free from court or government favouritism. Newton's contemporaries, he declared, would have been gratified to see how 'the desire of knowledge pervades the length and breadth of the land, and that there is no class of the community which does not more or less participate in it'.[37]

The Cambridge contingent made no such claims to hold a democratic view. One University representative rejoiced that only upper- and middle-class people attended; never before, he announced, had so many 'respectable' people (his inverted commas) crowded the streets of Grantham. Unsurprisingly, Whewell used these Grantham celebrations to reinforce Cambridge's hegemonic rule over England's scientific kingdom. Although he condescendingly deigned to 'recognize the connection which other places may claim with him [Newton]', he stressed that Cambridge was Newton's intellectual birthplace. Trinity men, he tactlessly bragged, shared a timeless academic kinship with their prestigious scholarly ancestors: 'In our cloisters, and our groves, where they paced to and fro weaving their speculations, we still pace to and fro, trying to solve the difficulties which their speculations invoke. We feel as if Newton had passed from among us only yesterday.'[38]

These tensions between promoting Newton as a local and a national hero still survive. In 1987, the 300th anniversary of Newton's *Principia* was celebrated internationally, but was also a Lincolnshire event. Hotels enticed visitors with 'Apple to

Astronaut' weekend breaks, and Grantham tourist agents took a sapling to New York – or as the local paper put it, the little apple visited the Big Apple. The 'Gravitate to Grantham' day included a carnival parade, a raft race, and an attempt to beat the world conga record. Sober national newspapers sneered at such provincial frivolity, but for Lincolnshire inhabitants, Newton carries iconic significance as a local hero. With no pretence of aiming at authenticity, they dressed up in an anachronistic assortment of historical costumes. For them – as for many other people – Newton has become a legendary figure who inhabits England's mythical past.[39]

Under a new policy of commercialization, in 1998 the National Trust rented Woolsthorpe to 'custodial tenants' who are committed to making it a Newton educational centre. Hundreds of Newton's collateral descendants attended a 'Newton Family Day', which was sponsored by Newton Investments Ltd and advertised by the slogan 'May the "Force" be with you'. Souvenirs included the predictable wooden apples, as well as tea-towels showing the 'family home' and Isaac Newton woolly dolls. The old barn was packed with modern Newtons, who eagerly forged links with relatives from all over the country by exchanging notes about a gigantic family tree displayed on the walls. Whewell may have regarded Newton as his intellectual kin, but these self-taught genealogical experts claimed genuine blood ties with this icon of Englishness.

London: institutional icon

Although Newton lived in London for the last thirty years of his life, for almost three centuries virtually the only outdoor public monument to him was an unobtrusive bust (now with a broken nose) in Leicester Square. Unlike those of Cambridge and Grantham, London's tributes to Newton were mostly related to his work rather than his homes. Because London is Britain's capital, the nation's major institutions are also located there, and they have laid claim to different aspects of his memory. The Royal Mint, for example, honours Newton not for his science but for saving the nation's money through his concern with accuracy. (Unfortunately, this punctilious approach apparently never seeped through to the design department: the 1978 pound note showed an incorrect diagram of Newton's planetary system, and a revised version three years later rectified one mistake but introduced another.)[40]

The chief London guardian of Newton's memory is the Royal Society, where his bust greets visitors in the entrance hall and several original portraits look down from the walls. Over the years, the Society's Library has acquired a treasured collection of personal relics of Newton – including a death-mask and a lock of hair – as well as instruments, manuscripts and books. But while Newton may be the Society's most revered former Fellow, its elegant mansion is more like a cathedral for a scientific community than a shrine to Newton the individual.

Newton moved several times while he lived in London, so enthusiasts have confusingly labelled at least three buildings 'Newton's home'. Throughout the nineteenth century, activists tried unsuccessfully to ensure some sort of memorial for his long stay in London. Campaigners wrote angry letters to newspapers, published nostalgic itineraries tracing his footsteps through the city, and wrote poems hymning the capital's great intellectual citizen. In addition to routine pleas for heritage plaques, more exciting suggestions included encasing Newton's house in a dome, and building an underground shrine to be opened up every fifty years. A more realistic approach was taken by souvenir hunters who plundered the Kensington house in which he had died. Subsequently taken over by a railway company, it was demolished in the 1890s, and the only memorial is a window in a nearby church.[41]

Newton lived longest in St Martin's Street, adjacent to what is now the National Gallery. This house's most famous later resident was Fanny Burney, who liked to absorb inspiration by writing in Newton's converted Observatory on the top floor.

Other inhabitants were less distinguished. One of them was a Frenchman, uncharitably said to have conned gullible visitors into believing that a collection of second-hand scientific instruments had belonged to Newton himself. This enterprising curator was almost certainly Claude-Mammès Pahin-Champlain de la Blancherie, who fled to England at the beginning of the French Revolution and embarked on a new career as a Newtonian publicity agent. Berating the English for neglecting their own hero, he solicited money to restore Newton's Observatory and set up a permanent memorial to 'this divine Person'.[42]

For years the increasingly dilapidated building was run as the Hotel Newton, but by 1852 Newton's Observatory had become a cobblers' workshop. After the house became a mission refuge for prostitutes, optimistic salvationists unrealistically claimed that 'Sojourning where great minds have been, will often cause the humble and lowly to aspire to such goodness and greatness.' Nevertheless, Newton's former home was demolished in 1913 by the London County Council, which managed to sell parts of it to an antiques dealer, who kept it in storage.[43]

American private collectors did more to preserve Newton's house than British national institutions. American bibliophiles had started to build up huge libraries of European books in the eighteenth century, when original copies of Newton's *Principia* first crossed the Atlantic. Despite English protests, the rate of acquisitions rapidly increased. The New England investor Roger Babson, who claimed to have gained a fortune by applying Newton's third law of motion to the stock market, amassed over 1,000 books by and about Newton, as well as pictures, his bed and a scion of the Woolsthorpe apple tree.[44]

When Babson's wife Grace visited England in 1937, she acquired their largest piece yet: Newton's fore-parlour from the house in St Martin's Street, purchased from the foresighted antiques dealer. Four years earlier, public concern about Britain's heritage being shipped to America had ensured that Captain Cook's cottage remained within the Empire, and it is now in Melbourne. However, Grace Babson gained permission to export

Newton's room and reconstruct it in rural Massachusetts. Visitors to Babson Business College can now experience a disconcerting transition as they step from a modern lift straight into an early eighteenth-century study. Grace Babson had aimed to re-create the feeling of Newton's home, but, as in many commercial heritage showplaces, some of the decorations are anachronistic. The original wood panelling and marble fireplace have been complemented by new curtains in deep crimson (known to be Newton's favourite and traditionally the colour of nobility), as well as other Newtonian memorabilia, including busts and pictures created after he died. This room that claims historical authenticity functions curiously as a preserved shrine that is also a miniature museum.[45]

Revealing his strong acquisitive urge, Babson preached that there are only three important days in a person's life: when they choose their ancestors, their career and their spouse. Babson was a firm believer in the power of heredity, and could trace his own family's arrival back to the *Mayflower*. Newton's parlour gave Babson the intellectual and aristocratic roots denied him by the genes in which he placed so much importance. While English scientists celebrated Newton as the founding father of modern science, Babson claimed him as the father of his investment fortune. He even gave his autobiography a Newtonian title: *Actions and Reactions*. Newton offers an ideal subject for a fetishistic collector. He is a scientific hero whose qualities can be appropriated through accumulating his possessions, yet there is no risk that the collector's nightmare of a completed collection can ever be realized.[46]

9

INHERITORS

> *'Work the furnace, Humphrey,' Newton growled. Humphrey had seen him like this before, going days without eating or sleeping, utterly consumed by thoughts that even other scholars could only vaguely guess at. If Isaac were merely deluded, Humphrey would not stand here pumping the bellows like a slave, but Newton was not insane. He was that rarest of creatures. He was a genius.*
>
> J. Gregory Keyes, *Newton's Cannon* (1998)

In 1835, the London astronomer and former stockbroker Francis Baily fired such damning accusations against Newton that the acrimonious rejoinders by prominent academics were covered in the national press. Baily claimed to provide documentary evidence that Newton had treated John Flamsteed, the Astronomer Royal, with almost criminally arrogant behaviour. Far more than historical accuracy was at stake. One of Whewell's Oxford allies protested, 'If Newton's character is lowered, the character of England is lowered and the cause of religion is injured.'[1]

Like many of his other quarrels, Newton's conflict with Flamsteed had lasted a long time. Their mutual vengefulness was barely concealed beneath a polite veneer. For instance, during their apparently friendly correspondence about Flamsteed's crippling headaches, Flansteed diplomatically declined Newton's solicitous (or vindictive?) recommendation to 'bind his head strait with a garter till ye crown of his head nummed'.[2] Their protracted contest about who held the rights over Flamsteed's observations at Greenwich was far from being a simple head-on confrontation between two obstinate and unpleasant men. Several of their colleagues – notably Hooke and Halley – were also intimately involved, and this tussle for the ownership of astronomical data paralleled the struggles of Pope and his fellow authors who were fighting to protect their literary works from plagiarization.

During his stockbroking career, Baily had – like Newton – remorselessly prosecuted illicit financial dealers, and now he was determined to defend Flamsteed and cleanse the scientific world of unethical practitioners. In accounts favouring Newton's point of view, Flamsteed had emerged as a secretive, lazy and bad-tempered man who repeatedly delayed publication of the figures that it was his job to provide for Newton. These versions implicitly assumed that theoretical work is more important in science than data collection, but Baily wanted to place methodical, precise observation at the heart of the nineteenth-century scientific enterprise.

Armed with Flamsteed's handwritten transcript of events, Baily set out to vindicate the Astronomer Royal's behaviour. According to his interpretation, Newton not only illicitly published the star catalogue that Flamsteed had compiled at his own expense, but also drastically altered its contents and constantly harassed the elderly astronomer. Now it was Newton who appeared sly and uncooperative. Baily depicted him as a bitter and reclusive theoretician whose intuitive speculations meant little without Flamsteed's systematic measurements. By redeeming Flamsteed at the expense of Newton, Baily had valued accurate, industrious data collection above the inspired insights of a scientific genius. He had used his research into this personal antagonism of the past to challenge the contemporary scientific hierarchy, in which the mathematical theoreticians at Oxbridge looked down on outsiders like himself, yet depended on the meticulous observations provided by dedicated astronomers.

Safeguarding Newton's prestige continued to be a matter of vital importance for patriotic scientists. Thirty years later, a heated international controversy erupted when a French astronomer claimed to own manuscripts proving that it was not Newton, but the French mathematician Blaise Pascal, who had discovered the laws of gravitation. *The Times* protested vigorously: 'Such an allegation touches our national pride . . . In the brilliant annals of scientific research the name of NEWTON shines with a matchless lustre . . . there is not an educated Englishman who does not feel a personal interest in his reputation.' It soon became apparent that hundreds of counterfeit letters had been successfully palmed off by an inept forger who had not even got his dates straight. Nevertheless, British experts continued to ferret out counter-evidence, and the ever-protective Brewster even asked the Royal Society to set up a special committee to protect his idol's character against this foreign assault.[3]

The first challenge to Newton's status as Britain's greatest scientific hero came from Charles Darwin. Often dubbed the Newton of biology, Darwin was – against his wishes – buried in Westminster Abbey close to Newton. British chauvinists crowed that two scientists had 'stamped their name on periods marked by the triumph of their theories: and only two. Both were Englishmen.' Nevertheless, they concluded, Newton was indubitably the greater genius. In 1927, at the 300th anniversary of Newton's death, his national standing seemed unshakeable. Patriotic scientists urged their countrymen to 'rejoice that this greatest of all men of science was British', boasting about his incomparable mental acumen as well as his satisfyingly British characteristics of modesty, 'all-roundness' and practical skill. This universally acclaimed genius was also the country's hero, fit to rival France's Napoleon or Italy's Michelangelo.[4]

Newton may have regarded himself as a giant who stood on others' shoulders, but new contenders for the position of outstanding genius would, in their turn, come to surmount him. During the twentieth century, the main competitors for Newton's place were Einstein and Hawking. In an episode of *Star Trek: The Next Generation*, this intellectual trio chats amiably together over a game of poker. Hawking – who plays himself – replays the scene on major publicity occasions, including his visit to the US President. As the complex constellation of science, society and genius continues to evolve, the real-life 'next generation' of geniuses may well be computer experts. In Newton's time, the 'Man of *Genius*' was held in far higher esteem than the 'Man of *Business* . . . the dull, plodding Mind [that] fixes all its Attention on the single Point of growing rich'. But creative genius is no longer divorced from financial ambition. As electronic systems mushroom in importance, the world's next great 'Man of *Genius*' may well be Bill Gates, that far from dull and plodding 'Man of *Business*' who – like Newton – became a legendary figure in his own lifetime.[5]

During the twentieth century, Newton was remembered and reinterpreted in new ways. The first threat to his supremacy came from Einstein, who found it advantageous to present himself as Newton's opponent in a head-on confrontation between two intellectual giants. As Einstein and his colleagues consolidated his reputation, their manoeuvres affected not only the meanings of Newtonianism, but also the characteristics of genius. The second major attack was inspired by political interests, when dramatic Marxist reinterpretations of Newton's activities provoked British historians into defending their national hero and reappraising the aims of scientific research. Yet a third onslaught on Newton's conventional image was launched by the economist John Maynard Keynes, who shocked the world by insisting that Newton was not the first great scientist, but the last great magician.

As the years between us and Newton increase, he is becoming more than ever a legendary figure. His achievements are so remote that he no longer provides an accessible role model for physics students, yet his symbolic value survives unscathed. In 1999, the British government embarked on a drastic sale of its property. The National Physical Laboratory sold its buildings to a private purchaser, but the garden, which boasted an apple tree grown from a Woolsthorpe cutting, was not on offer. Even for a depleted Treasury, it seems that Newtonian national assets are just too valuable to lose.[6]

Like others who have passed into mythical realms, there can

be no end to the Newtonian image-making process. Dalí's sculpture (*Figure 0.1*), with which this book opened, illustrates this point. At one level, it is immediately astonishing that a Spanish surrealist should choose to make several statues of Newton. More symbolically, Dalí's hollow figure forcefully conveys that Newton has become an emblematic hero, leaving admirers who know nothing about his actual life or science free to provide their own interpretations. New rituals are constantly developing to commemorate versions of Newton that never existed in reality, but which evoke conflicting visions of Britain's heritage and of scientific genius.

Albert Einstein: a revolutionary genius?

Just as the falling apple became a legendary tale epitomizing Newton's achievements, so too the heroic story of Newton's own downfall became one of the fundamental myths surrounding Einstein's life. In an international radio broadcast, George Bernard Shaw hailed him as Newton's successor who had proved that 'Newton's theory of the apple was wrong.' Emphasizing Newton's Englishness, Shaw declared that Newton had created a wonderful universe 'and established it as a religion which was devoutly believed for 300 years . . . [until] the whole Newtonian universe crumpled up and was succeeded by the Einstein universe'.[7]

In November 1919, newspapers throughout the world announced that a young professor from Berlin had unexpectedly triumphed over Newton by producing a daring new theory of the universe. With varying degrees of clarity, reporters explained how a 'revolution in science' meant that conventional ideas about space and time must be replaced by the exciting if incomprehensible postulates of general relativity. Einstein became internationally famous almost overnight, yet beforehand this 'epoch-making' experiment had seemed so insignificant that the *New York Times* had sent along their golfing correspondent to cover the story. Sporting and military metaphors converted an academic debate about abstruse cosmological concepts into a thrilling personal contest between two scientific champions.

In England, Cambridge physicists were immediately deluged by worried inquiries about this apparent defeat of Newton, hero of both the University and the nation. Ironically, the supposedly decisive observations stemmed from the findings of two British teams, recently sent abroad to record a total solar eclipse. Despite the Great War, planning had started two years earlier. Patriotic tensions exacerbated the suspicion with which many British scientists regarded Einstein and his theories, but one enthusiastic supporter was the prominent astronomer Arthur Eddington, who managed to avoid detention as a conscientious objector by leading an eclipse expedition to Africa.

According to Einstein, a ray of light travelling from a star towards the earth is bent if it passes very close to the sun, so that the star appears to be in a slightly different place from usual. Although Newton's theories on their own have nothing specific to say on this subject, Eddington converted the eclipse measurements into a direct confrontation between Newton and Einstein by claiming that a small apparent shift in a star's position would correspond to the Newtonian system, while a larger one would corroborate Einstein's predictions.

Einstein was so convinced of the truth of general relativity that he regarded these tests as almost superfluous. Reprimanding one eminent physicist for staying up all night to hear the results, Einstein slept tranquilly, confident of the outcome. Eddington was also committed to obtaining the results he wanted. Problems with the weather and the optical instruments plagued both expeditions, but by carefully sifting through the ambiguous data, Eddington managed to salvage enough confirmatory evidence to declare triumphantly that Einstein was indisputably right.

Previously little known outside scientific circles, Einstein transcended wartime hostilities to become acclaimed all over the world as a scientific genius rivalling Newton himself. Within only a couple of years, he was being showered with honours and portrayed as 'the new Columbus of science voyaging alone through the strange seas of thought' – a deliberate reference to Wordsworth's poetic description of Newton.

Einstein readily cooperated with demands to popularize his ideas, writing books as well as accepting countless invitations to give public lectures and attend fashionable dinner parties. An adept self-publicist, he supplied punchy aphorisms for journalists eager to explain relativity. 'An hour sitting with a pretty girl on a park bench passes like a minute,' he quipped, 'but a minute sitting on a hot stove seems like an hour.'[8]

Like Newton's *Principia* or Hawking's *A Brief History of Time*, Einstein's relativity became a fashionable topic of conversation amongst people who had little real understanding of the basic concepts. Einstein's idiosyncratic behaviour and appearance lent themselves to parody, and fuelled his identification as a genius. As he became a cult figure, artists, writers and musicians created resonances – sometimes rather spurious ones – between their own innovations and his revolutionary scientific changes. Through photographs and cartoons, Einstein's shaggy hair, droopy moustache and rumpled clothes became instantly recognizable trademarks of this intellectual hero who – like Newton before him – embodied the power of reason.[9]

Einstein, like Newton, has become a legendary character whose achievements have acquired a mythical aura. Modern scientific historians deliberately echo Newton's *annus mirabilis* in Woolsthorpe by describing 1905 as Einstein's own miraculous year, when he published five extraordinary papers on a range of topics, including relativity. But just as the universe was not transformed overnight when Newton watched the apple fall, so too, Einstein's articles were extremely influential but did not immediately give birth to modern physics. Accounts of instantaneous transformation are deceptive. Although he now enjoys singular fame, Einstein was just one of many researchers who contributed to the major changes in scientific understanding during the first decades of the twentieth century. Even the development of relativity theory, the topic nowadays uniquely linked with his name, spanned a long period and owed much to the work of other physicists and mathematicians. Far from being an instant success, Einstein's ideas were hotly disputed and frequently modified, while some experimental tests of his general relativity theory that he suggested in 1915 were still being implemented in the 1960s.

Bertrand Russell, an early expositor of Einstein's relativity, explained to his readers that even Newton himself 'was not a strict Newtonian'.[10] The theories and practices of so-called Newtonian science were very different from each other as well as being far removed from Newton's own. Scientists had moved away from Newton's focus on the forces between particles, and increasingly couched their theories in terms of energy. Starting from a Newtonian base, they introduced new ways of thinking about the world, such as electrical and magnetic fields radiating throughout space, the efficiency of engines and the cooling of the earth. Only a few years before Einstein's *annus mirabilis*, scientists had been congratulating themselves on successfully elucidating the laws of nature. Future research would, they confidently judged, be a simple matter of tidying up the last few decimal places. The discovery of radioactivity abruptly shocked them out of this complacency, and it seemed that only disturbingly counter-intuitive theories could explain this new sub-atomic world.

Major revisions of Newtonian thought that had originated in the late eighteenth century developed in divergent directions. By the time that Einstein became a student, ways of explaining the world differed to the point of incompatibility, and the style of science you practised depended on where you lived. Like nations annexing colonies, commented one physicist, ambitious scientific communities were trying to extend their power. To Einstein's colleagues, it seemed that presenting him as the lone inventor of relativity could enhance the reputation of the departments where he worked. Glossing over the divergences between different approaches to the universe, propagandists bracketed them together as classical Newtonian science. This strategic move enabled them to convert Einstein into the sole author of a revolutionary theory that had overturned the world of physics.[11]

Einstein himself staked his claim more cautiously. Often paying tribute to Newton's genius, throughout his life he insisted that 'No one must think that Newton's great creation can be overthrown in any real sense by this or by any other theory . . . what we have gained up till now would have been impossible without Newton's clear system.' This modest reverence for an intellectual ancestor brought advantages. By stressing that science is a collective, progressive enterprise, Einstein placed himself in an elite international community transcending barriers of time. He eulogized a Newton who dwelt in the 'happy childhood of science', thus subtly implying that he was Newton's natural inheritor in the more complex world of modern adult science. For Einstein, Newton was an artist who delighted in his creative powers. His repeated adulation suggests that he too wished to be bracketed with Galileo and Newton, 'the greatest creative geniuses . . . whom I regard in a certain sense as forming a unity'.[12]

Through highlighting his affiliation with Newton, Einstein implicitly distinguished his own ideas from those of his peers, and so contributed to forging his identity as the sole originator of relativity. From his earliest papers in 1905, Einstein and his contemporaries contrived to make the introduction of relativity a highly significant event. By emphasizing the theory's novelty and impact, they reinforced their authority within their own institutions as well as amongst the international scientific community. By the time that Eddington set himself up as referee in a direct experimental confrontation between Newton and Einstein, theoreticians had already appointed Newton to be Einstein's superseded opponent.

At the crossroads of science

Until the end of June 1931, British academics remained serenely unaware of another powerful attack that was about to be unleashed on Newton's reputation. The International Congress of the History of Science and Technology may not appear a promising site for a revolution, yet some scenes that took place in London that summer sound as if they have been taken from a David Lodge novel. In one session, tempers became so heated that the chairman resorted to silencing a Soviet denunciation of Newton hero-worship by ringing a ship's bell.[13]

At that time, history of science scarcely existed as an academic subject, and this Congress was only the second to be held. The discipline was so young that scholars had not yet coined the term 'Scientific Revolution', now a commonplace for describing the changes that took place between around 1550 and 1700. Science's past was seen as continuous progress, a march towards truth in which men like Galileo and Newton handed on the torch of knowledge to their successors. According to the ideology of the early twentieth century, scientists operated in the realm of pure ideas and left mundane cares at the doors of their laboratories. This cosy model was about to be disrupted.

Only a few days before the conference was due to begin, Stalin surprised the Soviet Union as well as Britain by announcing the imminent arrival of a large delegation in London. Headed by Nikolai Bukharin, one of Lenin's former advisers, the Soviets demanded that the schedule be rearranged to accommodate all their hastily prepared talks. The British organizers adamantly refused to cancel a day trip to the Newtonian shrine of Trinity College, but in compensation, within less than a fortnight the Soviet papers had been translated into English and published together as *Science at the Crossroads*. Astonishingly, the book was promptly favourably reviewed in the *Spectator*, a prominent conservative weekly.

The Soviets' searing critiques of Western capitalism baffled and embarrassed most of the other delegates. Seventy years on, they make strange reading – dense, rhetorical lectures in which the concepts of dialectical materialism have been rendered into an English that is grammatically impeccable but terminologically alien. In retrospect, it was the contribution by Boris Hessen,

Director of the Moscow Institute of Physics, that proved the most influential.

'The social and economic roots of Newton's *Principia*': even Hessen's title seemed sacrilegious to most British listeners. Hessen poured scorn on the tradition of attributing progress to the personalities of individual great men. Deliberately focusing on Newton, the example *par excellence* of the historical approach he wanted to discredit, Hessen argued that the *Principia* was not the brain-child of an isolated scholar, but the product of seventeenth-century class struggle. As the feudal system disintegrated in the face of rising merchant capitalism, he explained, the path of scientific research was determined by the urgent demands of the emerging bourgeoisie for better communication systems, military equipment and industrial processes. Just as Newton had helped to create a new science for a new political age, so too, under a socialist regime, science would advance victoriously for the benefit of all humanity.

With brutal clarity, Hessen stated his intention to demolish the prevailing image of 'Newton as an Olympian standing high above all the "earthly" technical and economic interests of his time, and soaring only in the empyrean of abstract thought'. Hessen's analysis was far from simplistic. Subtly exploring the philosophical and theological attitudes of the period, he knowledgeably compared Newton's concepts of space, time and matter with those of his contemporaries. Hessen portrayed an entirely new Newton, one who was deeply immersed in the technical and economic issues of his time. Thus he studied alchemy to improve metallurgy and coin manufacture, mathematics to help soldiers fire cannon more accurately, and hydrodynamics to design new canals and mine pumps.[14]

Like Bukharin, Hessen disappeared during the era of Stalinist repression, but his erudite Marxist study prompted radically changed accounts of the relationships between science and society, and played a vital role in the emergence of history of science as an academic discipline. Hessen's provocative reappraisal of Newton's *Principia* generated immense antagonism, but also converted eminent scientists. One of the earliest was Joseph Needham, now celebrated for his massive studies of Chinese science and technology, but then a practising embryologist. Needham urged future researchers 'to do for the great embryologists what has been done so well by Hessen for Isaac Newton'.[15]

Following Hessen's polemical lecture, some startling new versions of Newton appeared. Drawing strength from Hessen's claim that capitalist entrepreneurs had inspired the scientific changes of the seventeenth century, British left-wing campaigners insisted that society would be revolutionized not by working-class activities but through the advancement of science. The most prominent spokesman for this view was the Marxist biochemist J. D. Bernal, who strongly influenced the social organization of science in Britain well into the 1960s. Deeply affected by the 1931 conference, Bernal made Newton and his peers not so much the inheritors of an intellectual dynasty extending back to the Greeks, as the beneficiaries of a recent bourgeois upsurge in trade and industry. According to him, Newton had a disastrously stultifying effect on science, but was 'a forerunner of the French Revolution' whose individualistic model had its greatest immediate impact on economic and political affairs.[16]

Less ideologically committed academics also paid tribute to Hessen's analysis. For instance, although stopping a long way short of Bernal's position, the American sociologist Robert Merton painted a very different picture of the seventeenth century from the traditional one. In a book that became a seminal text for historians of science, Merton substantially downplayed Newton's status as a singular man operating solely on an intellectual plane. His Newton was just one particularly outstanding member of a large group of talented men who turned their attention towards scientific matters to meet the contemporary demands for technological improvement.[17]

The focus of these authors on ballistics, gunpowder and other military inventions stemmed from their interest in the relationships between science and warfare. Writing in the late 1930s, Bernal urged scientists to recognize that the Great War and the economic depression should oblige them to appraise the social function of science. Committed to organizing scientific research for society's benefit, Bernal especially deplored Hitler's regime for harnessing science to Germany's military requirements. He quoted with repugnance the anti-Semitic article written for *Nature* by Johannes Stark, the Nazi scientist who denounced relativity and quantum mechanics as Jewish physics. For Stark, Newton was a prime example of a Nordic pragmatic thinker, so evidently different from Jewish dogmatists. Ironically, arguing from the opposite side, Bernal drew much the same contrast, stressing that Newton was an experimenter who displayed a characteristically English practical and common-sense approach to science. Since Englishmen are incapable of thinking systematically, claimed Bernal, the only hope for the country's scientific future lay in receiving greater financial support from the state, and in assimilating Jewish immigrants with their powerful theoretical skills.[18]

Despite these radical reappraisals of Newton's contributions, his ideological role as a detached genius survived. Bernal's dream of government-funded science was unfortunately realized: because of the Second World War, money was channelled into military development, precisely the application he had tried to eliminate. In the 1950s, horrified by the atomic bomb, scientists were forced to confront the consequences of their research. One solution was to make science a purely theoretical endeavour. Einstein, who had been one of Stark's targets, recommended turning to Newton for solace. Perhaps thinking of how he had himself been accused of encouraging the American bomb programme, Einstein used a military metaphor to portray Newton 'as a scene on which the struggle for eternal truth took place', one of those creative intellects automatically absolved from responsibility for misusing scientific knowledge.[19]

During the Cold War, when people were trying to make sense of recent upheavals by placing them within a longer historical context, history of science gained strength as a specialized topic within the natural sciences. Determinedly anti-Marxist scientific historians stressed that 'history is made by men, not by causes or forces'. Like Einstein, they emphasized Newton's innocent genius, sneering at the notion that his 'sublime, impersonal science' might have any connection with political or social thought. 'Science . . . has the independence of any work of art,' they protested, and 'tells what we can do, never what we should'.[20]

In 1952, A. Rupert Hall, one of England's earliest professional historians of science, directly contradicted Hessen's claims that military requirements had directed scientific research. Hall restored the priority of abstract knowledge over technology. According to him, men like Newton had 'used the problems of ballistics as a gymnasium in which to develop their powers for larger and more important researches'. Only in the nineteenth century, he insisted, had manufacturers turned for help to this independent science that had been created 200 years earlier.[21]

Hall's mentor at Cambridge had been Herbert Butterfield, a history professor whose radio broadcasts had comforted huge post-war audiences by locating the true meaning of history in Christianity. Uniquely among religions, argued Butterfield, Christianity's ethic of tolerance enabled people to accept new ideas while still retaining their spiritual outlook. For him, there had been only two outstanding events in human history. The first was the rise of Christianity, and the second was the Scientific Revolution of the seventeenth century, when the publication of Newton's *Principia* provided an internationally recognized landmark that held deep significance for societies throughout the world. Whereas civilized ideals had formerly been transmitted by Christianity, now it was science that carried Western styles of thought to every corner of the globe. The Christian religion had evolved into a new secular faith of science, with the *Principia* as its biblical text.[22]

Butterfield's profoundly Christian interpretation continues to dominate modern narratives of science's history. Forty years later, Hall had become world famous as a Newton expert, and he published his own biography of Newton as an 'adventurer in thought', a being who was almost pure mind, and to whom 'sensual and aesthetic experiences were denied'. Pronouncing that 'thinking man is as much the hero of the *Principia* as sinful man is of *Paradise Lost*', Hall made Newton an intellectual Adam in a scientific Bible.[23] If the Scientific Revolution represents the mythical birth of a new stage in human progress, then Newton must be its scientific saviour. Perhaps that why so many people remark that Newton's 1689 portrait, painted only two years after he completed the *Principia*, resembles an image of Christ Himself (*Figure 2.1*).

Modern icons

Newton was born on Christmas Day, 1642. This date now seems extraordinarily appropriate for a secular saint, but its significance was scarcely commented on by eighteenth-century writers. The Puritans had abolished Christmas observance in the mid-seventeenth century, and it was only during Victorian times that Christmas started to take precedence over Easter and become commercialized. England's belated conversion to the Gregorian calendar added further confusion, since in France and some other European countries, Newton's date of birth was 4 January 1643. Even in the twentieth century, some critics found it sacrilegious to conflate Newton's birthday with a religious festival. But by the tercentenary of his birth, this fortuitous date provided a powerful image for Maynard Keynes, who portrayed Newton as a Christ-like figure, 'a posthumous child born with no father on Christmas Day, 1642 . . . the last wonder-child to whom the Magi could do sincere and appropriate homage'.[24]

The year 1642 also saw Galileo's death, but because of the complicated calendar changes, this only became significant in the middle of the nineteenth century. Victorians used this happy correspondence of Newton's birth and Galileo's death to corroborate the heritability of genius. One popular poet flamboyantly exclaimed, 'On thee his robe the parting prophet flung.' Even in sober journals, writers used this same image of exchanging garments to invest genius with religious, transcendent characteristics: 'The mantle of the Tuscan sage seems scarcely to have dropped from his shoulders, when a mightier spirit arose to receive the garment, and to take office as the Interpreter of the Heavens.'[25] Hawking exploits his own birth in 1942, 300 years later, to reinforce his image of being Galileo's natural successor as well as the inheritor of Newton's Lucasian Chair.

This emphasis on precise personal milestones is surprisingly recent. For example, that enterprising promoter David Garrick let Shakespeare's bicentenary pass by unmarked. It was only in the second half of the nineteenth century that, fuelled by an increasing interest in the past, the cult of the centenary gathered strength. The British tourist industry benefited from the huge popularity of pageants and concerts, but academic historians grumbled about the frequent disruptions of their normal life, complaining that 'We shall soon have as many centenarized heroes . . . as canonized saints.' During the Newtonian ceremonies of 1927, journalists moaned that 'celebrations are now of almost monthly occurrence, and some people may think there are too many of them.' They were unaware how rapidly the frequency of centenary festivities would accelerate and become commercialized, so that now it is hard to recognize that they are a relatively new innovation.[26]

The term 'celebration' was originally reserved for religious occasions. Slowly its meaning expanded first to embrace royal ceremonies, and then to describe more mundane occasions such as birthday parties. In Hanoverian England, secular heroes like Newton assumed some of the cultural attributes of kings, earlier believed to be appointed by God as His representatives on earth and the possessors of special therapeutic powers. Voltaire remarked with amazement that Newton was buried like royalty, because in France it was still unthinkable to honour a commoner, however distinguished, with such splendid funeral rites. It was only during the nineteenth century that death became democratized, and ordinary people adopted ceremonial practices that had previously been the preserve of the privileged classes.[27]

Rituals are traditionally associated with religious observance, but members of secular communities are also bonded together by sharing in activities that symbolize their beliefs about how the world is ordered. Repeatedly performed, such rituals establish a continuity with the past, but they also constantly evolve over time as participants invest them with new significance. Scientific anniversary celebrations have become increasingly secular, money-making events, which emphasize a heroic ancestor's importance but also reinforce the significance of modern science. They are uneasily riven by the tension between reiterating the scientific myth of progress towards absolute truth, and the inherent necessity of acknowledging that earlier convictions have been discarded. These shifts reflect the changing status of science as well as wider transformations in our attitudes towards remembering the past.[28]

In the modern Newtonian calendar, major festivals are associated with his birth (1642), death (1727), and the publication of the *Principia* (1687). Before the twentieth century, all these significant dates seem to have passed unnoticed. Although the French émigré Pahin-Champlain de la Blancherie did campaign to restart the historical clock from 1642, little attention was paid to his manifesto of 154 (more conventionally known as 1796).[29] Proving a historical absence is, of course, impossible. How should we interpret the 1787 publication of *Principia Botanica* by the doctor Erasmus Darwin, Charles's grandfather? Did he simply select a title with strong Newtonian resonances, or did he deliberately choose this significant date, exactly 100 years after the appearance of Newton's own *Principia*? Although it seems strange to us, the lack of any contemporary references to a Newtonian anniversary suggests that his apparently neat timing was, in fact, coincidental. A century later, the Fellows of the Royal Society commemorated the fiftieth anniversary of Queen Victoria's coronation, but passed over the 200th anniversary of the *Principia*.

Although 1787 and 1887 seem to have passed unnoticed, in 1987 numerous celebrations throughout the world marked the tercentenary of Newton's *Principia*. Professional historians and scientists assembled for international conferences, those ritual occasions when participants adopt formal clothes for an official dinner, and an elder of the clan gives a keynote address valued for its symbolic value rather than its keen insights. At Grantham, in a bizarre re-enactment of a mythical event, Hawking promoted himself as Newton's natural successor by ensuring that he was photographed in the Woolsthorpe garden, sitting beneath a supposed descendant of the original apple tree. For academic audiences, he tackled the conflict between triumphal pictures of progressive science and the need to reject older beliefs. Presenting himself in relation to Newton and Einstein, he portrayed modern cosmology as a comprehensive system that did not replace his predecessors' models, but instead accommodated them as special cases.[30]

The 1987 anniversary celebrations of the *Principia* far outshone those commemorating Newton's birth and death. An increasingly high premium has come to be placed on individual creativity, as illustrated by the virulence with which Shakespeare experts argue over their hero's authorship. Newton's own book has acquired the status of a scriptural text. In 1942, one enthusiast suggested that 'the tercentenary of his birth might be celebrated by a re-reading of *Principia*', conceding only that a translation might be substituted for the original Latin. By 1998, a first edition of the *Principia* fetched nearly £2 million in a New York auction house. Singling out the *Principia* as a unique text distorts the past by consigning all its predecessors to oblivion and also obliterating the individuals who contributed to its production and dissemination. Similarly, Newton's apple paradoxically condenses the birth of physics, which relies so heavily on mathematical reasoning, collective observation and institutional research, into a flash of inspired theoretical insight in a secluded country garden.[31]

Intellectual capital is now invested in publications, the quantified measure of academic achievement for individuals as well as for institutions. Compared with Newton's contemporaries, we invest relatively little significance in the place where an idea originated. The most recent monument to Newton, Paolozzi's statue (*Figure 1.2*), is located not in a site imbued with the spirit of Newton's personal genius, but in the forecourt of the British Library. Like the Library itself, this statue raised academic hackles and was the victim of governmental cutbacks. Originally intended to be accompanied by other British intellectual heroes, Newton now sits in solitary splendour, a national scientific figurehead who owes his bronze existence to commercial sponsorship by a football pools consortium.

Horrified by Paolozzi's deliberate references to Blake's image of Newton (*Figure 6.1*), critics wrote angry letters to the press accusing the Library of indulging in ridicule. In private, lavatorial comparisons flourished, compounded by rumours that the Library had taken revenge on authoritarian planning officials by orienting Newton so that his back pointed towards Camden Town Hall. Paolozzi and the architects hotly defended his reinterpretation of Newton, comparing it with Auguste Rodin's *Thinker* and insisting that it presented a subtle yet Michelangelesque vision of science's ambiguous role. Blake contrasted a mechanical Newton, perched on a rock, with his organic aqueous surroundings, but this robotic bronze statue is seated on a hard geometric pedestal amidst a sea of square paving slabs. Paolozzi's Newton may be less enigmatic than Blake's, yet it evokes modern technoscience by radiating power through its design as well as its massive size.[32]

Newton's prominent position in one of London's new cultural showpieces indicates his significance as a national hero, and also the centrality of science to modern notions of scholarship. Yet while the *Principia* may have earned him this prestigious spot in an academic temple, it is his falling apple that endears him to the world. The British heritage industry has helped to convert Newton and his apple into commercial products. Grantham tourist literature invites visitors to drive out to the Sir Isaac Newton pub, which disappointingly turns out to be not an idyllic thatched cottage, but a characterless modern building whose sign shows a bemused Newton about to be hit on the head by an apple. Wooden apples and apple pies entice tourists at Woolsthorpe,

Cambridge's Newton Institute displays his 'favourite pudding' recipe, and Web browsers can choose between an expensive apple made of cherry wood ($129 in 2000), a replica of Newton's 'very unique' walking stick, and an oddly labelled Newtonian fountain pen (the first fountain pen patent was not granted until 1884).[33]

Like the wood of the Cross, offspring purporting to be from the original tree flourish in biological research stations and university gardens all over the world. Their authenticity is now being tested by radiocarbon dating and genetic fingerprinting, expensive procedures that are presumably being paid for from national science budgets. The prominence of Newton's apple in his mythology can only have been helped by chauvinistic campaigns to defend British varieties against European imports. Ironically, as environmental concerns become more pressing and also more fashionable, this former emblem of nature has become central to high technology. When Apple computers ceased producing their Newton range, newspapers punned liberally on the falling apple image: the headline above a cartoon very similar to Leech's (*Figure 7.1*) ran: 'Users get the pip as Apple knocks Newton on the head'.[34]

It was not until 1970 that the Bank of England started to make the currency an agency of nationalism by illustrating great heroes. Although the Post Office chose a large apple as its Newtonian symbol for its 1987 commemorative stamps, the Mint opted for the more subtle reference of a flowering tree on its pound note. This is no longer in circulation, but when the new £2 coin appeared near the end of the century, the phrase 'standing on the shoulders of giants' was inscribed around the rim. This unique style of tribute seems particularly suitable, since Newton's major innovation was to insist that coins are milled to prevent clipping.[35]

The humble researcher who claimed to have stood on the shoulders of giants remains a potent image, even though Newton the dog-lover and pipe-smoker have disappeared. This version of Newton advertises science as a progressive, continuous project, whose participants belong to a community that transcends barriers of time and is rapidly expanding. As one physicist quipped, since the vast majority of all the world's scientists are alive today, they 'are uniquely privileged to sit side-by-side with the giants on whose shoulders we stand'. The expression has entered public awareness so deeply that it became the title of an Oasis CD – well, almost. Because of an unfortunate error after a few drinks in a pub, it is rather strangely called *Standing on the Shoulder of Giants*.[36]

The alchemy of genius

Newton's scientific reputation may now be secure, but his character is constantly being reappraised. Many biographers are still haunted by the same allegations of corruption that troubled the elderly Voltaire, who confessed that he had become disillusioned with his idol:

> I thought in my youth that Newton made his fortune by his merit. I supposed that the Court and the city of London named him Master of the Mint by acclamation. No such thing. Isaac Newton had a very charming niece, Madame Conduitt, who made a conquest of the minister Halifax. Fluxions and gravitation would have been of no use without a pretty niece.[37]

Was Newton indeed bribed to accept an adulterous relationship between his beautiful niece Catherine Barton and the influential statesman with an amorous reputation, Lord Halifax? The evidence is so complicated that historical sleuths have drawn differing conclusions, which reveal much about their own moral stance. Many of Newton's admirers have found it hard to believe that Newton could have condoned an illicit affair, which must have been happening while Barton was living in Newton's house and looking after him. Nevertheless, it remains hard to provide any other explanation for Halifax's extraordinarily generous legacy to Barton – or la Bartica, as she was dubbed by contemporary gossipy journalists.

Victorian scientists were scandalized: Brewster denied the very possibility, while Augustus de Morgan wrote a whole book to exonerate his hero by postulating a secret marriage. As social codes relaxed in the later twentieth century, historians accommodated Newton's apparent culpability by divorcing his moral conduct from his intellectual achievements. Evidently working hard at positive thinking, Richard Westfall transformed Newton's lapse into proof that he was no 'plaster saint', but a human being forced to pay the price for his genius. Still more creatively, Frank Manuel suggested a Freudian explanation, cautiously framing it as a question: 'In the act of fornication between his friend Halifax and his niece was Newton vicariously having carnal intercourse with his mother?'[38]

Newton died almost three centuries ago, yet new visions of his life are constantly being created. Like a hologram, his image flickers back and forth between two apparently irreconcilable models. On the one hand, he is a paragon of detached rationality who represents modern ideals of scientific thought. By 1927 Newton had come 'very near to Nietszche's description of "the objective man", a passionless being concerned only to "reflect" such things as he is tuned to perceive'.[39] But this Nietzschean Newton has gained an increasingly prominent double, the mad genius, the solitary recluse verging on insanity who seeks the truths of nature in alchemical crucibles, arcane symbolism and cabalistic calculations.

John Maynard Keynes, assiduous collector of Newton's neglected alchemical manuscripts, initiated this transformation just after the Second World War, although he died before he could appreciate the impact of his words. 'Newton', he declared, 'was not the first of the age of reason. He was the last of the magicians, the last of the Babylonians and Sumerians, the last great mind which looked out on the visible and intellectual world with the same eyes as those who began to build our intellectual inheritance rather less than 10,000 years ago.'[40]

According to the biographer Michael White, it was not until the very end of the twentieth century that the 'real' Newton emerged. Yet White's reclusive, ill-tempered alchemist who distils reason in his alembic flasks is neither more nor less true than the nineteenth-century accounts of a dedicated, patient, methodical thinker who lived a pure and holy life. Today's Newton is an unpleasant, self-preoccupied introvert who is obsessed with alchemical experimentation, but at the same time he remains an icon of rationality.[41] This duality as both an insane genius and a dispassionate scientist is unique to Newton. Writers may fondly recount Einstein's eccentric refusal to wear socks, or sympathetically analyse Darwin's secretive, semi-invalid lifestyle, but among scientific icons, only Newton is simultaneously mad, bad and brilliant.

In the late 1990s, racy novels featuring Newton deliberately blurred the boundaries between fact and fiction. These pseudo-historical books are deceptively packed with authentic details, yet it is often hard to discern where reality slides into fantasy. Newton's alchemical and sexual activities provide the major focus of these narratives. Under the guise of literary fantasy, modern authors have gained the freedom to explore precisely those aspects of Newton's character that Victorian moralists had tried to suppress. In *Newton's Niece*, Newton's homosexuality and voyeurism drive a complicated plot of alchemical transformation and criminal corruption. The fictionalized Bartica gazes round her bedroom and realizes 'the true significance of my uncle's choice of decor. He wasn't just the civil servant he claimed to be. He had a new project . . . a whorehouse – a laboratory whorehouse – and I was the whore.'[42]

Newton the brothel-keeper was also voted Man of the Millennium, an accolade that might have surprised even the most loyal of his contemporaries. Selection juries engaged in long behind-the-scenes deliberations to reconcile conflicting choices. Newton's achievements, scoffed the member of one panel, were 'puny' by comparison with the innovations introduced in the previous millennium by Plato, Aristotle and Jesus; his own favoured candidate was Murasaki Shikibu, the eleventh-century inventor of the novel. More conventional nominations included Martin Luther, William Shakespeare and Karl Marx, but there was overwhelming agreement that a scientist would best reflect the most profound type of change that had taken place during the previous 1,000 years. Should this be Darwin, Einstein or Galileo? Eventually a history professor's terse judgement won the day: 'Newton. End of story!'[43]

But Newton's story has no end . . .

Notes

In the Notes, works listed in the Bibliography are referred to by author's name only, or by author's name and short title when more than one work by the same author appears in the Bibliography, or when works by different authors with the same name are cited.

1: Sanctity

1. *Times* index, *passim*; *Times*, 20 July 1857, 6c, 6 January 1866, 12a, 23 March 1886; *Illustrated London News*, 2 January 1864; *Personal Computer World* (May 1998), 27.
2. *Universal Magazine* 3 (1748), 295 (the Marquis de l'Hôpital).
3. The fullest and best biography of Newton is Westfall, *Never at Rest*, my major source of information throughout this book, to which I shall refer only for quotations. There are far too many other modern biographies to list here, but the ones mentioned in this book include: Hall, *Newton: Adventurer in Thought*; Manuel, *Portrait of Newton*; White, *Last Sorcerer*. Fauvel *et al.* has some excellent introductory essays, and Gjertsen is useful for reference.
4. Isaac Barrow, quoted in Westfall, *Never at Rest*, p. 102.
5. Quoted *ibid.*, p. 143.
6. Humphrey Newton (no relation), quoted *ibid.*, p. 209.
7. Bate, *Shakespeare*; Dobson; Holderness.
8. Poole. By the time of Newton's death, the difference had grown to eleven days.
9. Williams, *Keywords*; Mann; Smith, *Four Words*; Yeo, *Encyclopaedic Visions*, pp. 146–55.
10. Kant, *Critique of Judgement*, §47, pp. 176–7; Schaffer, 'Genius'.
11. Abrams; Bone.
12. Quoted in Iliffe, '"Is he like other men?"', p. 176; Osler.

13. Woodward; Brown.
14. Johns, p. 320; Snobelen, 'Reading Newton's *Principia*' (letter to William Derham quoted p. 159).
15. British Library Add. MSS 32548, fol. 32 and 32456, fol. 73.
16. Williams, *Pope*; Jenkyns; Webster, 'Taste'; Maty, pp. 62–3.
17. Desaguliers, *Experimental Philosophy*, vol. 1, p. vi; Shapin, 'Of gods and kings'; Stewart, 'Other centres of calculation', pp. 133–4.
18. Young, p. 76; Jordanova, 'Science and nationhood'; Anderson, pp. 11–49.
19. Daniels, especially pp. 32–42; Roe, pp. 33–45.
20. *London Journal*, 26 August 1732 (edited version in *Gentleman's Magazine* 2 (1732), 917–18) (I am grateful to Mark Goldie for this reference).
21. *Gentleman's Magazine* 8 (1738), 591 (reprinted from *Universal Spectator*, 25 November 1738); Mullan.
22. *London Journal*, 26 August 1732 (edited version in *Gentleman's Magazine* 2 (1732), 917–18) (I am grateful to Mark Goldie for this reference).
23. Albury. Halley's poem was edited several times, and the lines on Motte's frontispiece are not identical to the original.
24. Walters; Nurmi; Essick. See also the frontispiece of vol. 2 of James Hervey's *Meditations and Contemplations* (1748 edition) and plate 10 of Blake's *There Is No Natural Religion*.
25. *Monthly Review* 20 (1759), 300–1 (on William Lovett, who showed Hutchinsonian and Behmenist leanings); Schaffer, 'Newtonianism'.
26. Quoted in Gascoigne, p. 234.
27. Gascoigne: Richard Bentley (Boyle Lecture of 1692), quoted p. 223.
28. Iliffe, 'A "connected system"'; Keynes.
29. Westfall, *Never at Rest*, p. xi, and 'Newton and his biographer'.
30. By Pyio Rattansi, Betty Jo Teeter Dobbs, Michael White.
31. Rob Iliffe, quoted in *IC Matters* (Autumn 1999), 13.

2: Icons

1. Maude, *Viator*, p. ii (Appendix). There are no complete catalogues of Newton's portraits and their engravings. The best sources are: Smith, 'Portraits'; Webber, pp. 203–21; Gjertsen, pp. 440–8; the archives of London's National Portrait Gallery.

2. Kneller painted a similar version that was acquired by Caroline of Anspach and kept at Hampton Court: Stewart, *Kneller* (1971), p. 64, and Millar, vol. 1, pp. 27–8. Iliffe, '"Is he like other men?"' and 'Isaac Newton'; Shapin, 'The philosopher and the chicken'.
3. Crompton, 'Portraits of Newton', p. 3; Fortune and Warner, pp. 51–65.
4. Wolf, p. 349 (letter from James Logan to William Burnett, 1727); see also p. 350; Martin, *Biographia Philosophica*, p. 361.
5. Simoni. For example, James Northcote's *The Worthies of England* and James Barry's *Elysium*.
6. Richardson, vol. 1, last page of unpaginated preface.
7. Atterbury, vol. 1, p. 180.
8. Pointon, *Hanging the Head*; Jordanova, *Defining Features;* Shawe-Taylor, *The Georgians* and *Genial Company*; Simon.
9. Muirhead, p. 29; Nenadic.
10. From a 1712 letter to the *Spectator*, quoted in Stewart, *Kneller* (1983), p. 58.
11. Villamil, p. 13; Maude, *Viator*, p. vi; Avery, *David Le Marchand*, pp. 76–8, and 'Missing'.
12. Letter from Bentley to Newton of 20 October 1709: Turnbull, vol. 5, pp. 7–8; Fortune and Warner, pp. 50–65; Baker, 'Verrio and Thornhill'.
13. Letter from Cotes to Newton of 20 July 1712: Turnbull, vol. 5, pp. 315–16 (quotation p. 316). Letters from Bernoulli to Varignon of 20 February 1721 and from Varignon to Newton of 26 September 1721: Turnbull, vol. 7, p. 166 and *ibid.*, pp. 160–66.
14. Guerlac. Letter from Varignon to Newton of 17 November 1720: Turnbull, vol. 7, pp. 104–7 (quotation p. 105); see also Newton's letter to Varignon of 19 January 1721: Turnbull, vol. 7, pp. 119–23; Manuel, *Isaac Newton: Historian*, p. 383.
15. Exceptions include a few Grand Tour parodies, a Kneller portrait of Pope, and a drawing of Newton by William Hoare.
16. Stukeley, pp. 12–13, 85; letters from Stukeley to Conduitt of 15 July 1727 (Keynes MS 136) and 22 July 1727, quoted in Brewster, *Memoirs* (1855), vol. 2, p. 414. See also Stewart, *Kneller* (1983), pp. 71–2.
17. Spence, vol. 1, p. 350 (on Lord Pembroke); Pointon, *Hanging the Head*, pp. 53–78.

18. Hawkins, Franks and Grueber, vol. 2, pp. 469–73; Smith, *Portrait Medals*; Snelling, plate 29.
19. From Addison's 1713 *Dialogues upon Medals* and Pope's epistle *To Mr Addison*, both quoted in Wimsatt, *Portraits of Pope*, p. 50.
20. Abbé de Guasco, 1767, quoted Haskell, p. 5; Yarrington.
21. Paulson, *Hogarth: High Art and Low*, pp. 1–4, 59, 100; Webb, 'Busts'; Uglow, pp. 168–70.
22. Bindman and Baker, pp. 9–23, 187–9.
23. Webb, *Rysbrack*, pp. 76–91; Llewellyn.
24. Dobson, pp. 134–84; *Universal Magazine* 3 (1748), 249, and 34 (1764), 241; *Gentleman's Magazine* 1 (1731), 159–60, and 11 (1741), 548. Several readers contributed Latin translations of Pope's couplet: *Gentleman's Magazine* 11 (1741), 601, 663, and 18 (1748), 164.
25. *Gentleman's Magazine* 1 (1731), 169. See also *Gentleman's Magazine* 11 (1741), 663.
26. Paulson, *Hogarth's Graphic Works*, pp. 183–5; Ralph, pp. 69–73; Physick, pp. 80–5.
27. *Royal Magazine* 9 (1763), 116–17, repeated in *Universal Magazine* 34 (1764), 241–2.
28. Berteloni Meli; Millar, vol. 1, pp. 27–8; *Universal Magazine* 3 (1748), 298.
29. *Pietas Academiæ*, E1 recto (by Edward Turner).
30. Hunter, pp. 126–8.
31. Webb, *Rysbrack*, pp. 146–54; Wilson, pp. 143–5, Webb, 'Busts'; Colton (quotation p. 918 from the *London Journal*, 1732).
32. Willis, pp. 106–27, plates 111–58; Hunt, 'Emblem and expressionism'; Etlin, pp. 184–97; Paulson, *Emblem and Expression*, pp. 19–34.
33. Hunt, *Figure in the Landscape*, pp. 127, 142; Gilpin, pp. 28–9.
34. Gilbert West, *Stowe*, reproduced in Hunt and Willis, pp. 215–27 (quotation p. 219).
35. McKitterick; Baker, 'Portrait sculpture'.
36. *Nature* (26 March 1927), 466.
37. Thomas and Ober, pp. 161–72.
38. Baker, 'Portrait sculpture'; Curtis, Funnell and Kalinsky; Montagu quoted in Bindman and Baker, p. 120.
39. Sharpe, p. 93.
40. Original by Bernard Picart, reproduced Haskell, p. 10; Duportal, p. 369.

41. McKendrick, Brewer and Plumb, especially pp. 100–45.
42. McSwiny; Haskell, pp. 4–13; *The European Fame of Isaac Newton*; McSwiny's 'Worthies' do not seem to be connected with those at Stowe, which were erected later.
43. Kinns, pp. 399–43.
44. Portrait by Joseph Highmore discussed in Kerslake; Symonds.
45. *London Daily Post* (1 May 1738); Birch.
46. *Universal Magazine* 3 (1748), 289–301.
47. Munby, *Cult of the Autograph Letter*.
48. Cambridge, pp. 328–9.
49. Curtis, Funnell and Kalinsky; Pierre Grosley, quoted in Bindman and Baker, p. 114.
50. Cooper, *Life*, pp. 103–4, 116–18; *Gentleman's Magazine* 11 (1741), 102; Coleridge quoted in Holmes, p. 265.
51. Thrale, vol. 2, p. 795. For Pope's portraits, see Wimsatt; Reily and Wimsatt.
52. D'Oench, pp. 45–6; Reily and Wimsatt, pp. 150–1.
53. Webb, *Rysbrack*, pp. 117, 221–2; Reily and Wimsatt, plate 11, pp. 145–6. The lines on the socle, adapted from Ovid's *Fasti*, were: *NEWTONUS ANGLUS: Promissum ille sibi voluit prænoscere calum / Nec novus ignotas hospes adire domos* (translation by David Money).
54. Webb, *Rysbrack*, pp. 197–9; Mallet, 'Portrait medallions'.
55. *Man at Hyde Park Corner*; Uglow, pp. 529–31.
56. Hughes, 'Portrait busts'; Dawson, pp. 64–86; Reilly; Taylor, 'Artists and *philosophes*'.
57. Henig, Scarisbrick and Whiting, pp. 281–3; *Times*, 16 September 1796, 3d (the gem was by Nathaniel Marchant, one of Tassie's French rivals); Gray, pp. 32–45; Raspe, pp. lxxv, 746, 799.
58. Halfpenny, pp. 139–40, 172–3; Hughes, 'Notable earthenware figures'; Rackham, vol. 1, p. 117.
59. Pye (there were 240 pennies in a pound).

3: Disciples

1. Shapin, 'Boyle and mathematics'.
2. Blake, p. 474 ('Annotations to Sir Joshua Reynolds' *Discourses*').
3. Nash (quotation p. 55); Cunningham; Osler.
4. Hume, p. 166.

5. Turnbull, vol. 1, p. 328 (letter to Henry Oldenburg of 5 December 1674); Snobelen, 'Reading Newton's *Principia*,' pp. 160, 159 (Gilbert Clerke).
6. Ditton: quotations from pp. 1–7 of unpaginated preface; see Markley, pp. 184, 213–14.
7. Cohen, *Science and the Founding Fathers*, pp. 97–134.
8. Addison, 'Oration,' pp. 204, 198.
9. Ll. 1–2 of Glover's unpaginated poem in Pemberton.
10. Maclaurin, pp. xix–xx (by Anne Maclaurin, referring to Cotes's preface of the second edition of the *Principia*); Yeo, *Encyclopaedic Visions*, pp. 146–55.
11. Quoted without a source in Nicolson, p. 17.
12. Guicciardini, pp. 262–4.
13. Mullan (quotations pp. 42–3).
14. Tollet, pp. 25–7; quotation pp. 66–7 (from 'Hypatia').
15. Algarotti (quotation vol. 1, pp. iv–v).
16. Stevens, pp. 68–9.
17. Schaffer, 'Newtonianism'; Hales (quotation p. xxxi).
18. *Guardian*, 24 March 2001, 11 (and elsewhere).
19. *Biographia Britannica*, vol. 5, pp. 3210–44.
20. Manuel, *Newton: Historian*, p. 142; Hunter, pp. 126–8.
21. Bond, vol. 1, p. 44 (12 March 1711).
22. My major sources for the following account are Force, and Stewart, *Rise of Public Science*, pp. 31–141.
23. Boyle's will quoted in Gjertson, p. 86.
24. Gascoigne; Alexander, p. 11 (letter to Princess Caroline of November 1715).
25. Snobelen, 'Newton, heretic' (Viscount Percival quoted p. 381).
26. Henry Newman, quoted in Force, p. 20; Snobelen, 'Reading Newton's *Principia*'.
27. Guicciardini, pp. 99–117; Dobbs.
28. Manuel, *Newton: Historian*.
29. Fara, pp. 134, 155–6.
30. Manuel, *Newton: Historian*, pp. 166–93; Popkin, 'Fundamentalism II'. For example, Priestley, *Chart of Biography*, p. 13, and Barbauld, p. 90.
31. Chambers, vol. 1, p. 399; Henry Fellows, quoted by Ada Lovelace in Toole, p. 99 (letter of 21 July 1837).
32. Cohen, *Science and the Founding Fathers*, pp. 97–134. The following

account is based on Popkin, 'Origins of fundamentalism' and 'Fundamentalism II'.

33. The Socinians were a related unorthodox religious sect. Henry Fellows, quoted by Ada Lovelace in Toole, p. 99 (letter of 21 July 1837); *The Great Mystery of Godliness Incontrovertible*, by Ebenezer Henderson (1830); Froom, vol. 2, pp. 653–69 ('Transmitting the luminous torch of prophetic interpretation' is the frontispiece of vol. 1).
34. Newton, *Daniel*, p. xiv. Jim Bramlett, at www.Idolphin.org/angels299.html, 14 Sepember 1999.
35. Clubbe, p. 12; Paulson, *Hogarth's Graphic Works*, pp. 196–7; Shell.
36. My major sources for this account are Bowles, Rousseau, and Roy Porter's introduction to Cheyne. Quotations from Cheyne, pp. 338, 326.
37. From *Essay on Regimen*, quoted in Bowles, p. 486.
38. Guerrini; Money, pp. 135–67.
39. Cheyne, p. 4.
40. Quotations from: letter to Samuel Richardson, quoted in Rousseau, p. 89; Cheyne, p. 327.
41. From *Philosophical Principles of Natural Religion* (1715 edition), quoted in Bowles, pp. 479–80.
42. Cantor and Hodge.
43. Popkin, 'Fundamentalism II', pp. 167–8. The major sources of the following account are Olson, pp. 236–83 (Priestley quoted on p. 238) and Halévy, pp. 5–34, 433–87.
44. Bichat, quoted in Hall, 'Biological analogs', p. 6.
45. Paulson, *Hogarth: Graphic Works*, pp. 52–4, 98–9, and *High Art and Low*, pp. 55–60, 99–101; Uglow, pp. 108–9, 167–70, 347–8. My major sources for this account are Rowbottom and Stewart, *Rise of Public Science*; for further references, see the *New DNB* article (by Fara).
46. Fara, pp. 31–65.
47. Desaguliers, *Course of Experimental Philosophy*, vol. 1, p. vii.
48. Desaguliers, *Newtonian System*, pp. 22–4.
49. Hales, p. 147.
50. Morton and Wess.
51. Desaguliers, *Course of Experimental Philosophy*, vol. 2, p. viii.

4: Enemies

1. Westfall, 'Fudge factor' (letter to Roger Cotes quoted p. 757).
2. William Chaloner, quoted in Westfall, *Never at Rest*, p. 574; Marquis de l'Hôpital, quoted *ibid.*, p. 473; Johns, especially pp. 444–621.
3. Maclaurin, p. 13.
4. Shapin, 'Hooke'.
5. Merton, *Shoulders of Giants*.
6. Dollond, pp. 289–90; Sorrenson.
7. Stevens, p. 134 (from 'The Demirep; or, I know who').
8. Emerson, pp. iv–v; Schaffer, 'Newtonianism'.
9. Berkeley, vol. 5 (quotation p. viii).
10. Cantor, *Analyst*; Benjamin.
11. Berkeley, vol. 5, p. 116 (§243).
12. Quoted Gascoigne, p. 223 (John Hancock, 1706 Boyle Lecture).
13. Yolton.
14. Fara, pp. 24–30, 210–12 (quotations pp. 24, 211).
15. William Bowman, quoted in *Times*, 1 February 1828, 4b. Exchange reproduced and discussed in Yolton, pp. 190–4.
16. Paulson, *Hogarth's Graphic Works*, vol. 1, p. 288.
17. Hutchinson, *Moses's Principia*, part 1, p. 30; Cantor, 'Revelation and Hutchinson'.
18. Cantor, 'Weighing light'.
19. Wilde, 'Hutchinsonianism'.
20. Williamson; Guest, pp. 123–240; Smart, pp. 64, 72, 79 (*Jubilate Agno*, B130, B195, B264).
21. John Ker of Kersland, quoted in Klopp, vol. 11, p. xxxvi. My major biographical source is Aiton.
22. Herschel, pp. 7–8.
23. The original is 'Les grands hommes ressemblent en cela aux femmes qui ne cèdent jamais leurs amants qu'avec le dernier chagrin et colère mortelle': letter from Princess Caroline to Leibniz of 24 April 1716, in Klopp, vol. 11, p. 91.
24. Rice, pp. 211–19; Galloway, pp. 7–12; Hawking, *Brief History of Time*, pp. 181–2. See Hall, *Philosophers at War*, and Shapin, 'Of gods and kings'.
25. Thackray (Archibald Pitcairne quoted p. 156).

26. Guicciardini.
27. As he saw it 'querelle entre M[r] Newton et moy, mais entre l'Allemagne et l'Angleterre': Kemble, p. 529 (letter from Leibniz to Princess Caroline of 10 May 1715). Klopp, vol. 11; Alexander; Brewer, pp. 25–8.
28. Bertoloni Meli. Letter from Leibniz to Princess Caroline in November 1715, reproduced in Alexander, pp. 11–12, and Klopp, vol. 11, pp. 54–5.
29. Freudenthal; Meyer, pp. 58–9.
30. Shapin, 'Of gods and kings' (Leibniz quoted p. 208).
31. Euler, vol. 2, pp. 38–42 (letter CXXV); Watkins.
32. Werrett (quotation from Michael Lomonosov).
33. Russell, 'Leibniz', pp. 365–6 (originally published in 1903).
34. Kant, 'What is enlightenment?' (quotation p. 17).
35. Hagner, pp. 309–11; Kretschmer, pp. 174–6.
36. Kant, *Anthropology*, p. 126.
37. Kant, *Critique of Judgement*, pp. 174–8 (§§46–7); Schaffer, 'Genius'.
38. Kistler; Abrams; Furst; Woodmansee; Ziolkowski.
39. Beddow; Krätz; Wells.
40. My major sources for the following account are: Burwick; Gage, *Colour and Culture*, pp. 107–15, 201–4, and *Colour and Meaning*, pp. 162–95; Sepper.
41. Matthaei, pp. 21–3.
42. Heffernan, pp. 140–5; Thomson, vol. 1, p. 256 ('A Poem Sacred to the Memory of Isaac Newton', ll. 102–11); see also vol. 1, p. 15 (*The Seasons*, ll. 228–37).
43. Boskamp, 'L'arc-en-ciel de Joseph-Marie Vien'.
44. Brewster, 'Review of Goethe', pp. 99, 131.
45. Dubos, pp. 45–66.
46. Abrams, pp. 303–12 (Keats quoted p. 307 (from *Lamia*)).
47. Haydon, *Diary*, vol. 2, pp. 55, 173; see also pp. 154, 190–1, 229, 262; Tallis.
48. Burwick, pp. 176–209 (Coleridge's letter to Thomas Pool of 1801 quoted pp. 177–8).
49. Thomas and Ober, pp. 244–54.

5: France

1. Hall, *Eighteenth-century Perspectives*, pp. 108–73. This chapter is informed by several basic texts, notably Ehrard, and Dhombres and Dhombres. Where available, I quote from published English translations, preferably contemporary ones; the other translations are my own, but I have reproduced the French originals in the notes.
2. Darnton, especially pp. 48–9, 71, 397 n. 32; Delisle de Sales, vol. 4, pp. 173–201; Malandian.
3. Diderot, *Indiscreet Jewels*, pp. ix–xlix, and *Oeuvres complètes*, vol. 3, pp. 1–290 (especially pp. 57–60, 130–4, 261–6); Vartanian.
4. *Universal Magazine* 5 (1749), 281. Jean Baptiste de Boyer d'Agens (1749), quoted in Olson, p. 111.
5. Brook Taylor (1718, to Rémond de Monmort), quoted in Baillon, p. 74.
6. Quoted in Taylor, *Sources of the Self*, p. 325.
7. Pearson, pp. 29–37.
8. D'Alembert, p. 81.
9. Jay, pp. 21–147.
10. D'Alembert, p. 47; Lemercier, 'Dédication' ('O LUMINEUX ESPRIT! . . . Toi, de qui l'œil sonda le sein de l'univers, / Grand ombre de Newton, je t'adresse mes vers!').
11. Place, vol. 3, pp. 130–1.
12. Voltaire, *Letters on England*, pp. 68, 57.
13. Gebelin and Morize, vol. 1, pp. 217–19, 231–3, 236–7; Hall, *Eighteenth-century Perspectives*, pp. 53–74 (Conduitt quoted p. 54).
14. Guerlac; Hall, 'Newton in France'.
15. Voltaire, *Letters on England*, p. 68.
16. Ricard, pp. 154, 296–7 ('Qui conduit ces mortels? Quelle étonnante audace / Leur fait ainsi franchir ces montagnes de glace, / Et braver ces torrens suspendus dans les airs . . . ? / . . . de plus beaux motifs ont enflammé leur cœur'); Iliffe, '"Aplatisseur du monde"' and Greenberg, *Problem of Earth's Shape*.
17. Du Châtelet, quoted in Guerlac p. 73; Sherlock, pp. 164–5.
18. Originally 'vaste & puissant génie, Minerve de la France': du Châtelet (includes poem from Voltaire's *Élémens*); Walters.
19. Zinsser, 'Translating Newton's *Principia*'.

20. Algarotti, vol. 2, p. 170; Zinsser, 'Émilie du Châtelet'; Terrall, 'Émilie du Châtelet' and 'Gendered spaces'.
21. Crow, pp. 1–22. The striking portrait of du Châtelet looking up from her mathematical studies is no longer atttributed to Latour, although he probably did paint her.
22. Bongie, pp. 148–62; Goodman (although Ferrrand is not mentioned); Condillac, *Philosophical Writings*, pp. 170–2; *L'Année Littéraire* 1 (1754), pp. 620–1 (see also p. 638); *Correspondance Littéraire*, 1 December 1754, p. 438.
23. Conisbee, pp. 11–42; Fried, pp. 109–11; Debrie; Nicholson; Nolhac and Walsh.
24. Boskamp, 'Mademoiselle Ferrand'; private communication from I. B. Cohen.
25. Schaffer, 'Halley, Delisle'.
26. 'M. de Voltaire parle enfin, & aussi-tôt Newton est entendu ou en voye de l'être; tout Paris retentit de Newton, tout Paris bégaye Newton, tout Paris étudie & apprend Newton': *Mémoires de Trévoux* 1740 (1956) and 1738 (1673–4).
27. Canning, pp. 142, 156; Dainville; Dundon; Murdoch, 'Newton's law of attraction'.
28. Evans.
29. Grieder; John Collett, *Grown Gentlemen Taught to Dance*, discussed in Leppert, pp. 82–4. Gudin de la Brenellerie, p. 61 ('Quoi? ces astres nouveaux sont vus dans Albion! / Dans ce pays célèbre et si fier de Newton! / Oh combien l'univers doit envier cet île!').
30. Rosenau. There were several engravings of medals by Roettier: see Smith, 'Portrait medals'. For advertisements, see *Mercure de France* (October 1735), 813 and (August 1766), 208; Yarrington; Ozouf.
31. Etlin; Charlton; McManners, pp. 33–67; Wiebenson, pp. 81–9.
32. Le Dantec and le Dantec, pp. 112–50 (quotation from Girardin p. 138); Girardin, pp. 33–40 (quotation p. 39: 'ces Génies privilégiés qui paroissent un instant pour honorer leur patrie & éclairer leurs semblables'); *Tour to Ermenonville*, pp. 77–82. The other three poets on the obelisk were Theocritus, Virgil and Gessner. The other pillars in the Temple des Philosophes commemorate Voltaire, Penn, Montesquieu and Rousseau.
33. Roucher, vol. 1, p. 317 ('Newton, sur l'aile du Génie, / Planoit, tenant en main le compas d'Uranie'): Murdoch, 'French muse'.

34. Saint-Lambert, pp. xxv–xxxi, 143, 163 ('s'éclairer entre Locke & Newton!'): de Nardis; Cameron.
35. Roucher, vol. 1, p. 81 ('Mais si-tôt que Newton, cet aigle audacieux / En face eût regardé le Roi brûlant des Cieux, / L'Homme brisa les fers de l'ignorance antique').
36. Genlis, vol. 1, p. 219.
37. *Observations sur la littérature moderne* 8 (1752), 5–6 ('Sans cet illustre François, qu'on doit regarder comme le Fondateur de la bonne Philosophie, la Grande Bretagne gémiroit encore sous la tyrannie des Péripatéticiens').
38. Meek (quotation p. 95, from *On Universal History*, first published 1808–11).
39. 'Descartes perçant les ténèbres de l'ignorance': Hinks; Bordes and Chevalier, pp. 166–7; Taylor, 'Artists and *philosophes*'; Pupil.
40. 'Newton découvre et montre la Vérité qui d'une main tient un prisme pour marquer la théorie des couleurs, et de l'autre un cercle aimanté pour désigner son système de l'attraction': Sorel, pp. 140–2; Vidal.
41. I am grateful to Anthony Turner for his comments on this 'instrument'.
42. Fontanes, vol. 1, p. 21 ('Newton, qui, de ce Dieu le plus digne interprète, / Montra par quelles lois se meut chaque planète, / Newton n'a vu pourtant qu'un coin de l'univers; / Les cieux, même après lui, d'un voile sont couverts').
43. Chênedollé, p. 11 ('Bientôt d'un jour plus vrai Newton frappa nos yeux: / Ce grand législateur des mondes et des Cieux . . . / l'Erreur fut détrônée, et dans l'immensité / Son compas porta l'ordre et la simplicité').
44. Dhombres.
45. Gudin de la Brenellerie, pp. 55, 57 ('L'audace des Français conçoit, hasarde, achève / Cent prodiges qu'ailleurs on prendrait pour un rêve. / Que la gloire leur parle, ils volent, et sa voix / D'un peuple efféminé fait un peuple intrépide'); see also pp. 138–40, 141–59.
46. Martin, *Lettres à Sophie*, p. 52 ('C'est Newton qui l'ordonne: à la voix du génie, / Les astres font entendre une douce harmonie; / Et l'immortalité, qui reconnait Newton, / Sur le front des soleils vole graver son nom').
47. Chénier, André, *Oeuvres*, pp. 130 (from *l'Invention*) ('Et qu'enfin Calliope, élève d'Uranie, / Montant sa lyre d'or sur un plus

noble ton. / En langage des Dieux fasse parler Newton!') and 403 (from *Hermès*) ('Mais ces soleils assis dans leur centre brûlant / Et chacun roi d'un monde autour de lui roulant / Ne gardent point eux-mêmes une immobile place . . . Un invincible poids / Les courbe sous le joug d'irrésistibles lois, / Dont le pouvoir sacré, nécessaire, inflexible, / Leur fait poursuivre à tous un centre invisible'). See also pp. 125, 557–9, 877–8; Smernoff (especially pp. 45–54); Dimoff (especially vol. 1, pp. 58–63, 101–2, 323 and vol. 2, pp. 38–74); Vidal.

48. Chénier, Marie-Josèphe, *Oeuvres*, vol. 7, p. 248 ('Qui . . . Rendit parfaits Virgile et Cicéron; / Ouvrit le ciel aux regards de Newton; / Le cœur humain à Racine, à Molière? / Je le répète: une exquise raison').

49. Chênedollé, pp. 24–5 ('Peuples! rassurez-vous: ces masses inféconde, / Dont vous avez tant craint le retour menaçant, / Ranimeront un jour le Soleil vieillissant. / Ainsi l'a dit Newton, et j'en crois son génie'); Schaffer, 'Halley, Delisle'.

50. Roucher, vol. 1, p. 318 ('Est maintenant semblable à ces sages Royaumes, / Où suffit une loi pour régir tous les hommes; / L'attraction: voilà la loi de l'univers'); Racine, p. 166 ('Exerçant l'un sur l'autre un mutuel empire, / Par les mêmes liens l'un & l'Autre s'attire / Tandis qu'au même instant & par les même loix / Vers un centre commun tous pesent à la fois'.)

51. Vidler, *Writing of the Walls*; Picon, pp. 256–334; Pérez-Gómez, pp. 129–61; Ziolowski, pp. 309–77.

52. Boullée, pp. 136–45 (quotation pp. 139–40) ('Être divin! . . . Si par l'étendue de tes lumières et la sublimité de ton génie, tu as déterminé la figure de la Terre, moi j'ai conçu le projet de t'envelopper de ta découverte'); Pérouse de Montclos, *Boullée* (1969) and *Boullée* (1994).

53. Pérouse de Montclos, 'De nova stelli'; Vidler, *Architectural Uncanny*, pp. 165–75; Vidler, *Ledoux*, pp. 267–76.

54. *Sistême Astronomique de la Révolution françoise*: private collection of I. B. Cohen; J. Houël (in Year 8), quoted in Rosenau, p. 116 ('Un globe, en tous les tems, n'est égal qu'à lui-même; / C'est de l'égalité le plus parfait embleme').

55. Starobinski, pp. 31–8, 59–79; Stafford, 'Science as fine art'; Charles Fourier, quoted in Beecher, p. 244; Boullée, pp. 137, 142 ('Temples de la mort, votre aspect doit glacer nos cœurs . . . C'était dans le

séjour de l'immortalité, c'était dans le ciel que je voulais placer Newton.')

6: Genius

1. McKillop, pp. 20–5 (Grove quoted from *Spectator* 635 on p. 23).
2. Mallet, 'Excursion', p. 694 (Canto II).
3. *Ibid.*
4. Pope, p. 68 (from 'An essay on criticism').
5. Pattison, pp. 164–6 (from 'To Mr Hedges, On Reading his *Latin* ODE to Dr. *Broxholme*').
6. Money, *Gentleman's Magazine* 11 (1741), 548, 641, 663.
7. Voltaire, *Letters on the English*, p. 112; Williams, *Pope*; Jenkyns.
8. Home, vol. 1, pp. 292–301 (quotation p. 299); Spadafora (quotation from John Clarke (1731) p. 49).
9. Spencer, 'Lucretius'; earlier partial translations existed; Albury; MacPike, pp. 203–9 (quotation from 1755 translation by 'Eugenio').
10. Lucretius, p. 99 (by Thomas Creech); Book III, ll. 1042–5 ('Ipse Epicurus obiit, decurso lumine vitae; / Qui genus humanum ingenio superavit, et omnis / Restinxit, stellas exortus uti aërius sol').
11. Pemberton, p. 2 of unpaginated poem by Richard Glover.
12. Thomas and Ober, pp. 17–20.
13. *Oxford English Dictionary*. *Ingenium* is the nominative of *ingenio*, which could imply either the nature or the use of his mind; *genus* means race, and is, like genius and genitals, derived from the verb *gignere*, to beget; Ditton, p. 3 of unpaginated preface; Pope, p. 261 (from 'Epistle to Richard Boyle, Earl of Burlington'): Fumaroli; Smith, *Four Words*, pp. 22–48.
14. Turnbull, vol. 1, p. 1415 (letter of 20 August 1669 to John Collins); letter of 16 August 1680 to John Sharp, quoted in Manuel, *Portrait of Newton*, p. 107; Martin, *Biographia Philosophica*, p. 362.
15. *The Occasional Paper* 3/x (1719), p. 5.
16. *The Prelude*, Book III, ll. 60–3; Thomas and Ober.
17. Quoted in Woodmansee, pp. 429–30.
18. Ault, pp. 1–4; Gage, *Colour and Meaning*, pp. 144–52; Essick; Greenberg.
19. Tallis; Blake, p. 818 (from an 1802 letter to Thomas Butts).

20. Greenberg, 'Blake's science'; Blake, p. 470 (annotation to Discourse 6).
21. Hayley, pp. 314–15; Romney, pp. 228–9, 235–7; Ward and Roberts, vol. 2, pp. 197, 201; Schaffer, 'Scientific Discoveries'.
22. Young, *Conjectures*, pp. 26–7; Abrams; Furst; Smith, *Four Words*, Schaffer, 'Genius'.
23. DeNora and Mehan; Dubos and Dubos, pp. 45–66; Becker.
24. Blake, p. 472 (annotation to Reynolds's Discourse 7); Shawe-Taylor, 'Genial company', p. 70.
25. Priestley, *History of Electricity*, p. 576; Schaffer, 'Priestley'; Reid (quotations pp. 12, 15, 23); Laudan.
26. Kriz, pp. 1–8 (quotation p. 3 from *London Packet*); Mann.
27. Holmes: Davy quoted p. 119, Hazlitt quoted p. 240 (and p. 545).
28. Klein, pp. 47–8 (quotation p. 48).
29. Wollstonecraft, p. 119; Alaya; Battersby; Nochlin, pp. 145–78.
30. Brewer, pp. 573–612 (quotation p. 580).
31. Thrale, vol. 2, pp. 795–6.
32. Aikin, pp. 139–45 (quotation p. 144).
33. Quoted in Brewer, p. 150.
34. Gerard, p. 14.
35. MacLeod, 'Paradoxes of patenting' and 'James Watt'; Miller (inscription p. 2).
36. *Champion*, 13 July 1742, reproduced in *Gentleman's Magazine* 12 (1742), 365.
37. By Thomas Wilson, in St Stephen's Walbrook, London.
38. Macaulay, p. 17.
39. George, no. 11941.
40. Rose, Brewer, pp. 125–66; Woodmansee.
41. Cobbett, vol. 17, pp. 999–1000 (1774); Macaulay, pp. 17–22 (quotation p. 22).
42. Miller (John Robison quoted p. 6).
43. Malthus, p. 49.
44. *New Monthly Magazine* 16 (1826), 32.
45. Gerard, especially pp. 317–434 (quotation p. 323); William Duff, also at Aberdeen, praised Newton in a long account of what he called 'philosophic genius'.
46. John Thelwall, *Monthly Magazine* 59 (1825), 401n. (about Spurzheim's claim to have invented phrenology).
47. Hazlitt, *Selected Writings*, p. 272 (from 'Originality').

48. *Ibid.*, pp. 257–62 ('Why the arts are not progressive') and 136–7 (from 'The Indian jugglers').
49. *Ibid.*, pp. 273, 276, and 272 (from 'Originality').
50. *L'Encyclopédie*, vol. 17, p. 630 (le Chevalier de Jancourt) ('l'Angleterre peut se glorifier, d'avoir produit le plus grand & le plus rare génie, qui ait jamais existé').
51. Lemercier ('Ma fable à ton génie est un hommage'); from *Claude Gueux*, quoted in Patterson, p. 242, n. 42 ('par une loi d'attraction irrésistible tous les cerveaux gravitaient . . . autour du cerveau rayonnant').
52. Dieckmann.
53. Mercier, *Mon bonnet de nuit*, vol. 1, p. 13 ('l'homme de génie, qui poursuivoit la vérité avec une sagacité si admirable'): Darnton, especially pp. 115–36, 300–36; Majewski; Patterson.
54. MacLeod, 'Paradoxes of patenting'; Hesse.
55. Mercier, *Le Génie*, pp. 25–6 ('Le Génie alors sous son nom, / Déploya sa grandeur divine; / Sous cent noms différens caressant la Raison, / Le Goût fut tour-à-tour *Molière, Fenelon* . . . / *Milton* souverain de l'Europe entier, / Le Génie inspira *Newton*, / Et *Pope*, & le sage *Addison* . . .'); Chénier, Marie-Josèphe, vol. 7, p. 248 ('C'est le bons sens, la raison qui fait tout . . . / Et le génie est la raison sublime'); Furst; Fumaroli; Jaffe; Cassirer, pp. 312–31.
56. Halévy, p. 19 (on reading Helvétius in 1769).
57. Hall, *Eighteenth-century Perspectives*, p. 65; Condillac, *Origin of Human Knowledge*, p. 288.
58. Manuel and Manuel, pp. 461–86 (quotation p. 470); Meek, pp. 41–118.
59. Manuel and Manuel, pp. 487–518 (quotation p. 492).
60. Condillac, *Origin of Human Knowledge*, p. 288; Condorcet, pp. 124–72, 196 (quotations p. 150).
61. Beecher, pp. 71–4.
62. Saint-Simon (quotation pp. 19–20) ('cette contrée qui a été constamment le refuge des hommes de génie').
63. Quoted Manuel and Manuel, p. 119 (1808); Manuel, *Saint-Simon*, especially pp. 59–129; Dhombres and Dhombres, pp. 302–13; Vidler, *Writing of the Walls*, pp. 83–102. Boullée produced his own designs in 1784 and 1785; the two relevant prize competitions were in 1785 and 1800; Saint-Simon was associated with the École Polytechnique during the 1790s.

64. Spencer, 'Fourier', pp. 19–20; Fourier, especially pp. xvi–xix, 3–16, 314–15; Beecher, especially pp. 71–4, 241–58 Lloyd-Jones; Vidler, *Writing of the Walls*, pp. 103–14.
65. Beecher, p. 66 and fig. 29.
66. Manuel and Manuel, pp. 717–34. Harrison, pp. v–viii, 615; Wright, *Religion of Humanity*.

7: Myths

1. *History Today* 21 (1971), advertisement for the *Financial Times* facing p. 1.
2. I am grateful to Somak Raychaudhury for this information.
3. Noyes, vol. 1 (*The Watchers of the Sky*), pp. v, 195–6; Searby, pp. 450–1; Williams, 'Passing on the torch'.
4. Brown; Woodward.
5. De Morgan, *Budget of Paradoxes*, vol. 1, pp. 126–7; A'Beckett, vol. 2, pp. 271–3; Paradis.
6. Barthes; Friedman and Donley.
7. Abir-Am, 'How scientists view heroes' and 'Historical ethnography'; Cajori; McNeil. The large literature on myth-making includes Connerton; Healy; Samuel; Wright, *Living in an Old Country*.
8. J. Secord, *Victorian Sensation*, pp. 515–32; Dunning-Davies.
9. Lightman; A. Secord, 'Science in the pub'; Jordanova, 'Science and nationhood'; Mali.
10. Linton, p. 83.
11. Stukeley, *Memoirs*, p. 20.
12. McKie and de Beer (a) and (b); Keesing; Wilbert.
13. *Gentleman's Magazine* 1 (1731), 169; Thomas and Ober, pp. 44–5; Genesis 4: 17–18 and 5: 21–4; Yeo, *Encyclopaedic Visions*, pp. 22–32, 125–41.
14. Bacon, vol. 4, pp. 20–1 (preface to the *Great Instauration*).
15. Thomson, vol. 1, p. 251 ('A poem sacred to the Memory of Isaac Newton', ll. 17–23: 'HAVE ye not listen'd while he bound the *Suns*, / And *Planets* to their Spheres! th'unequal Task / Of Human Kind till then. Oft had they roll'd / O'er erring man the Year, and oft disgrac'd / The Pride of Schools, before their Course was known / Full in its Causes and Effects to him / All-piercing Sage!').

16. Byron, Canto X, stanzas 1–2 (p. 375); Chandler, pp. 365–6.
17. Shapin, 'Philosopher and the chicken'; Iliffe, 'Isaac Newton'; More's portrait reproduced in Manuel, *Portrait*, facing p. 111 (and elsewhere); Schaffer, 'Earth's fertility', especially p. 130.
18. Thomas, pp. 192–223; Stafford, *Last of the Race*, pp. 109–33.
19. Hogg, vol. 1, p. 5; Dryden, 'The Second Book of the Georgics', ll. 595–9; Morrill, p. 291.
20. Chartres; Evelyn, Appendix. Letters between Henry Oldenburg and Newton reproduced in Turnbull: 18 January 1673 (vol. 1, pp. 255–6); 2 September 1676 (vol. 2, pp. 93–4); 24 October 1676 (vol. 2, p. 110); 14 November 1676 (vol. 2, p. 181).
21. *Mirror of Literature* 4 (1824), p. 399 (by Nemo); Maude, *Wensley-dale*, p. 29; Jerman, pp. 168–9.
22. Holmes, p. 130.
23. Watt; Ash; Poole, pp. 17–18.
24. Brewster, *Life*; Shortland and Yeo; Schaffer, 'Scientific discoveries'; Secord, *Victorian Sensation*, pp. 46–51.
25. Brewster, *Memoirs*; Christie; Yeo, 'Alphabetical lives'.
26. Cantor, 'Scientist as hero'; Vicinus.
27. Hall, *Eighteenth-century Perspectives*; Yeo, 'Alphabetical lives'.
28. Spence, vol. 1, p. 462; *Paradise Regained*, book IV, l. 330.
29. John Herschel, quoted in Schweber, p. 69; Byron, Canto VII, stanza 5 and Canto IX, stanza 18; Blau, p. 75.
30. Martin, *Biographia Philosophica*, p. 373; Merton, *Shoulders of Giants*.
31. *British Quarterly Review* (1855), 336–7; Haydon, *Diary*, vol. 4, pp. 94–5 (June 1833).
32. Maude, *Wensley-dale*, p. 28; Sir John Hobhouse, quoted in Haydon, *Diary*, vol. 4, pp. 125–6 (August 1833).
33. Brewster, *Life*, pp. 222–41; Brewster, *Memoirs*, vol. 2, pp. 138–9; Edleston, pp. lx–lxiii; Hazlitt, *Works*, vol. 8, p. 239 (*Table Talk*); Biot, 'Newton' (published in French in 1821); *Buds of Genius*, pp. 27–30; *Lives of Learned and Eminent Men*, pp. 159–75; Shaw.
34. John Conduitt quoted in Westfall, *Never at Rest*, p. 49.
35. Kris and Kurz; Nochlin, pp. 145–78; Cooper, *Cooper's Journal*, pp. 233–4; *Athenæum* (1882), 93–4.
36. Edgeworth and Edgeworth, p. 240: I am grateful to Marilyn Butler for this reference; Shelley, *Letters*, vol. 1, pp. 50–1 (letter to his father of 6 February 1811); Yeo, 'Genius', especially p. 273.
37. Spence, vol. 1, p. 245; Shaw.

38. *Edinburgh Review* 103 (1856), 526; see also Galloway; Ditchburn.
39. *Times*, 21 September 1855, 8e–9a.
40. Shapin, 'Philosopher and the chicken' (quotation p. 21).
41. Hitchcock, p. 666.
42. Turnbull, vol. 2, pp. 437, 441 (letters of 20 and 29 June, 1686); Voltaire, p. 70.
43. Bigg, p. 115; Jerman. See also *Blackwood's Edinburgh Magazine* 70 (1851), 2 and Cooper, *Cooper's Journal*, pp. 250–1. Paradis.
44. *Times*, 21 September 1855, 8e–9a (see also *Phrenological Journal* 18 (1845), 154, and *Christian Remembrancer* 31 (1856), 365).
45. *How We Used to Live*, Yorkshire TV, 24 January 1995.
46. Quoted in Iliffe, '"Is he like other men?"', p. 176; Stukeley, p. 57; Humphrey Newton (no relation), quoted in Manuel, *Portrait*, p. 105.
47. Whewell; Yeo, 'Genius' and *Defining Science*, pp. 116–44; Theerman.
48. *Quarterly Review* 110 (1861), 401.
49. Crompton (1866) (quotations pp. 3, 5) and Crompton (1867); Barlow; Becker; Yeo; 'Genius'; Theerman. In 1850, the frontispiece of Edlestone showed a similar drawing owned by Pepys.
50. Crompton (1866), p. 3; *Antiquary* 15 (1887), 104–7; *Leisure Hour* 1 (1852), 634 (the picture is not named, but was clearly either Vanderbank's or a derivative such as Seeman's).
51. Martin, p. 362; McKie and de Beer (a) and (b); Keesing. The first printed references to the story were in 1727, the year Newton died.
52. Biot, 'Newton'; Brewster, *Life*, p. 344, and *Memoirs*, vol. 2, pp. 416–17; Biot, 'Brewster's *Newton*', p. 265n.; Galloway, p. 10; Chandler, pp. 155–202.
53. Biot, 'Brewster's *Newton*', pp. 193–4; Outram; Yeo, *Defining Science*, pp. 135–8.
54. *Illustrated London News* 56 (1870), 589, 594; Schaffer, 'Glass works'.
55. Yeo, 'Scientific method'.
56. *British Quarterly Review* 22 (1855), 328.
57. *Illustrated London News*, 6 October 1860, 282, 310, 320 and 13 October 1860, 339, 344; O'Dwyer, pp. 252–7; Blau, pp. 48–81; Acland and Ruskin, pp. 25–8, 79–81, 102–4; Forgan.
58. Dellheim, pp. 1–31, 157–75.
59. Abrams, pp. 187–98 (quotation p. 192).
60. Ellison, p. 183 (from 'The correspondencies of nature'). I am grateful to Thomas Coke for suggesting the Methodist comparison.
61. *Quarterly Review* 110 (1861), 421.

62. Morrell and Thackray.
63. Carlyle, p. 443; Turner.
64. Gaskell, p. 75; De Morgan, *Budget of Paradoxes*, vol. 1, pp. 375–6.
65. *Phrenological Journal* 18 (1845), 153–6 (letter from C.P.); Vago, pp. 76–83; Inwards, p. 8.
66. Yeo, 'Idol'; *Edinburgh Review* 1 (1832), 29–37.
67. Williams, 'Passing on the torch'; Yeo, *Defining Science*, pp. 116–44.
68. De Morgan, 'Theory of probabilities', p. 242, *Essays on Newton*, pp. 18–23, *Budget of Paradoxes*, vol. 1, p. 137, *Penny Cyclopaedia* 19 (1841), 5–12 (arguing against Stephen Rigaud) and 16 (1840), 197–203); Rice.
69. Cooper, *Triumphs*, pp. 103–4, 141.
70. Cooper, *Purgatory of Suicides*, p. 337.
71. Horne, p. 129; Ellis, p. 227 (and p. 118).
72. Leighton, p. 141 (from 'Lowly work'); Smiles, p. 71. See also Cooper, *Triumphs*, p. 102 and *Cooper's Journal*, p. 235; Craik, vol. 1, pp. 1–7; Buckley, pp. 218–28; *All the Year Round*, 18 April 1868, 443–4.
73. *Saturday Magazine* 1 (1832), 13–14 and 7 (1835), 241–3; *Lives of Illustrious Men*, p. 73. Massey, p. 21. See also Cooper, *Paradise of Martyrs*, Book IV.
74. *Youth's Instructer* [*sic*] and *Guardian* 21 (1837), 78.
75. *Buds of Genius*, pp. 29–30; *Lives of Learned and Eminent Men*, p. 174. See also Chambers, vol. 1, p. 399, and Leighton, p. 141.
76. Jalland, pp. 17–58. For example, Halford, p. 278; Cooper, *Cooper's Journal*, p. 252.
77. Suzuki; personal communication from Timon Screech; Fauvel, p. 240.
78. From 'Letter to Alex Comfort', in Heath-Stubbs and Salman, pp. 296–7.
79. Haydon, *Lectures*, p. 56, and *Diary*, vol. 2, p. 229 (see also pp. 154, 190–1); Falkner, p. 11; Cooter.
80. Whitwell, p. 314; Carlyle, p. 449.
81. Milner, p. 38.
82. Brewster, *Life*, p. 13, and *Memoirs*, vol. 1, p. 20.
83. King, *Biographical Sketch*, pp. 77–99; Timbs, pp. 149–52.
84. *Athenæum* (1882), 93. See Telescope, pp. 211–14; Craik, vol. 1, pp. 1–7, 196–210, vol. 2, p. 115.
85. Hodgskin, pp. 76–99 (quotation p. 89); MacLeod, 'Concepts of invention'.

86. Telescope, p. 212; see also Edgar, p. 170; Smiles, p. 244; Combe, p. 130.
87. Secord, 'Newton in the nursery'.
88. Casteras; Nochlin, pp. 145–78.

8: Shrines

1. Disraeli, p. 120; Brown; Stukeley, p. 86.
2. *Nature* 119 (1927) 467 and supplement for 26 March; *Grantham Journal*, 26 March 1927, 4–8; *Times*, 19 March 1927, 14; Friedman and Donley.
3. Newspaper cuttings in Royal Society MS 657/item 11; Geary; Wattenberg; *Times*, 21 March 1727, 10; Inwards; Vago.
4. Munby, *History of Science and Cult of the Autograph Letter*; Smith, *Historical and Literary Curiosities*.
5. Bate, *Shakespeare*, pp. 82–8; Outram; Pointon, 'Shakespeare'.
6. The vast literature on these topics include Abir-Am, 'How scientists view heroes' and 'Historical ethnography', Bensaude-Vincent; Cohen, 'Commemorations', Connerton; Healy; Holderness; Lowenthal; McNeil; Nora; Samuel; Winter; Wright, *Living in an Old Country*.
7. Sotheby, pp. 7–10.
8. Yeo, *Defining Science*, especially pp. 110–13 (Mary Somerville quoted p. 112); Alaya.
9. *Times*, 29 April 1885, 9f, and 9 May 1892, 6e; 14 November 1890, 10e, and 29 November 1904, 86.
10. *University of Cambridge Annual Report 1998–9*, p. 26.
11. Winstanley, p. 434; Adrian; Edleston, p. xliv; McKitterick, p. 107; Ditchburn.
12. Jalland, pp. 265–83; King, *Newton*, p. 101; Yeo, *Defining Science* (Whewell quoted p. 16); Browne, pp. 180–8.
13. Yeo, 'Genius' and *Defining Science*, especially pp. 116–44; King, *Newton*, p. 101.
14. Prowett, pp. 236–7; Searby, pp. 423–544; Geary.
15. Edleston; Adrian.
16. Dening, pp. 75–111; Okri; Trinity College, Add. MS c.242.
17. Brewer, pp. 325–489; Bate, *Shakespeare*; Holderness; Keats quoted in Bate, *Keats*, p. 356 (letter of 11 July 1818).

18. Maude, *Viator*, p. vii (Appendix).
19. Spence, vol. 1, pp. 351–2; Maude, *Wensley-dale*, p. 30; Cooper, *Cooper's Journal*, p. 219. Quoted Brewster, *Life*, p. 344 (the same verse, though differently punctuated, appeared on the engraving of Woolsthorpe in Maude, *Wensley-dale*, facing p. 28).
20. Charles Burney, quoted in Brewer, p. 472; Anderson, pp. 11–49.
21. *Edinburgh Review* 78 (1843), 436; Jordanova, 'Science and nationhood'.
22. Turnor, pp. 157–86.
23. Biot, *Life of Newton*, p. 265n.; de Morgan, *Paradoxes*, vol. 1, p. 137; Gordon, p. 260.
24. Mais, p. 1224, Gaze, pp. 158–64; *Notes and Records of the Royal Society* 5 (1947), 34–6; Healy, pp. 30–41.
25. Cooper, *Life*, pp. 103–4, 116–18; *Stamford Mercury*, 29 September 1826, 4 (see also 27 July 1827, 3).
26. Moore; MacLeod, 'James Watt'.
27. *Quarterly Review* 11 (1861), 433–5; Cantor, 'Where the statue stood'; Brewster quoted in Gordon, p. 147 (from the *North British Review*).
28. *Art Journal* 4 (1858), 243, 372; Yarrington; Munsell; Barlow.
29. *Quarterly Review* 11 (1861) 435; unidentified newspaper article (evidently from 1853), Royal Society Box MS 657/9; *Times*, 16 September 1858, 12c, and 23 September 1858, 8c. Extract printed in *Times*, 25 September 1858, 10e. See also *London Journal* 28 (1858), 133–4.
30. *Times*, 21 January 1854, 6d; Royal Society Council Minutes, vol. 2, 250–1 (26 May 1853; see also MM.XIV.10) and 449 (28 October 1858); Royal Society Box MS 657/9.
31. Yarrington; Munsell; Jenkyns; Vago, pp. 76–83; Inwards, p. 8; *Phrenological Journal* 18 (1845), 153–6 (letter from C.P.).
32. *Illustrated London News*, 2 October 1858, 316; *London Journal* 28 (1858), 133. See also *Times*, 16 September 1858, 12c.
33. Acland and Ruskin, pp. 79–81.
34. *Times*, 22 September, 1858, 7b, and 23 September 1858; 8c; see also 9 September 1858; 10b, 10 September 1858, 7e. Brewster did not attend, probably because he was convalescing from bronchitis: Gordon, pp. 165–6.
35. Bell, pp. 91–169; *Illustrated London News*, 25 September 1858, 284 and 288, and 2 October 1858, 315–16.
36. King, *Newton*, p. 90.
37. White, *Journals*, p. 119; King, *Newton*, pp. 109–12 (Sir Benjamin Brodie quoted p. 111); Macleod, 'Whigs and savants'.

38. King, *Newton*, pp. 117, 100–6.
39. *Stamford Mercury*, 24 June 1988; *Times*, 18 April 1987, 10a; *Grantham Journal*, 10 July 1987; *Sunday Times*, 22 March 1987, 31a.
40. *Sir Isaac Newton*; Brackenridge, pp. 89–93; Blaazer.
41. Ellison, p. 99; Thomas Steele, *Times*, 23 June 1834 (see also *Times*, 25 June 1834, 2e, and Morrell and Thackray, p. 343); *Leisure Hour* 1 (1852), 634–7; Jopling; *Times*, 9 August 1873, 12c, and 3 August 1870, 10f; *European Magazine* 60 (1811), 281–3; *Notes and Queries* 167 (1934), 164–5, 223; *Times*, 26 April 1866, 9a and 27 April 1866, 11b.
42. Smith, *Antiquarian Ramble*, vol. 1, p. 24; Goodman, pp. 242–80; Pahin-Champlain de la Blancherie ('ce divin Personage').
43. Masters (quotation p. 170 from the *Hackney Carriage and Taxi-Cab Gazette* of 1909); Hartill; King, *Wonderful Things*, pp. 10–12.
44. Basbanes; Smith, 'Portraits'; Webber; Gardner, pp. 92–100.
45. Healy, p. 33; Babson.
46. Smith, *Yankee Genius*; Bal.

9: Inheritors

1. This account is based on Ashworth (quotation from Stephen Rigaud p. 215) and Johns, pp. 543–621.
2. Turnbull, vol. 4, pp. 152–3 (letters of 20 and 23 July 1695).
3. *Times*, 2 October 1867, 8d (in *The Times* alone, there were thirteen letters and articles on the subject in under two months); Farrer, pp. 202–14; correspondence between Brewster and Sabine of 10 and 31 October 1867, Royal Society MSS, MC.8.91.
4. *Nature*, 26 March 1927, supplement, p. 24 (F. S. Marvin); Schweber, pp. 70–1; Moore; Jeans (quotation p. 28) and other articles in the special supplement to *Nature* of 26 March 1927.
5. *Champion*, 13 July 1742, reproduced in *Gentleman's Magazine* 12 (1742), 364–6 (quotation p. 365). A series of American books by Gene Landrum links genius with computer expertise.
6. Traweek, pp. 74–105; *Independent on Sunday*, 24 October 1999, 6.
7. Quoted Brian, pp. 230–1 (in 1930). My major souces for the following account are: Brian, pp. 100–6; Earman and Glymour.
8. Calaprice: President Hibben of Princeton University (1921), quoted p. 234; Einstein, p. 184.

9. Friedman and Donley.
10. Russell, *Analysis of Matter*, p. 14.
11. Staley; Schaffer, 'Newtonianism'.
12. Einstein, pp. 58, 222; Newton, *Opticks*, p. vii; Calaprice, p. 74.
13. My major source for the following account is Werskey, pp. 138–49.
14. Hessen (quotation p. 24).
15. Needham, quoted in Werskey, p. 147.
16. Bernal, *Social Function*, p. 23, and *Science in History*, pp. 337–43.
17. Merton, *Science, Technology and Society* (first published 1938).
18. Bernal, *Social Function*, pp. 191–237.
19. Einstein, *Later Years*, pp. 219–23 (quotation p. 219).
20. Gillespie, pp. 117–57 (quotations pp. 151, 150, 154, 521).
21. Hall, *Ballistics*, p. 158.
22. Cabral; Butterfield, especially, pp. 160, 179 (first published 1949).
23. Hall, *Isaac Newton: Adventurer in Thought* (quotations p. xiv).
24. Poole, *Time's Alteration*; *Times*, 19 March 1927. 11f–12a (H. H. Turner) and 21 March 1927, 10 (letter); Keynes, p. 27.
25. Tupper, p. 134; *British Quarterly Review* 22 (1855), 321.
26. Quinault (1866 *Pall Mall Gazette* quoted p. 320); *Nature* 119 (1927), 467; Cohen, 'Commemorations'.
27. Llewellyn; Jalland.
28. Abir-Am 'Historical ethnogoraphy' and 'How scientists view heroes'; Bensaude-Vincent; Bell.
29. Pahin-Champlain de la Blancherie, p. 118.
30. *Grantham Journal*, 10 July 1987; Mialet, especially pp. 562–73; Hawking, 'Newton's *Principia*'.
31. Holderness; Bate, *Shakespeare*, pp. 65–100; Mais, p. 1225; Secord, *Victorian Sensation*, pp. 515–32.
32. Letters to *The Times* (1, 10, 13, 20 August 1992) and to *The Times Literary Supplement* (19 March and 9 April 1993).
33. http://www.levenger.com/Newton/
34. Keesing, pp. 387–90. *Personal Computer World* (May 1998), 27.
35. Blaazer; Brackenridge, pp. 89–93.
36. Gerald Holden (1961) quoted in Merton, p. 266; Melvyn Bragg and Norman Thrower have both titled books with this expression; *Independent*, 18 December 1999, 5.
37. Quoted Westfall, *Never at Rest*, p. 596 (from Voltaire's *Dictionnaire philosophique*).
38. Westfall, *Never at Rest*, p. 601; Manuel, p. 262.

39. *Times Literary Supplement*, 17 March 1927, 167.
40. Keynes, p. 27.
41. White, *Last Sorcerer*, p. 3; *The Six Experiments that Changed the World*, Channel 4, 12 March 2000.
42. Beaven, p. 76; Keyes.
43. *Sunday Times Magazine* (12 September 1999), 40 (David Starkey).

Bibliography

A Tour to Ermenonville (London: for T. Beckett, 1785).

A'Beckett, Gilbert Abbott. *The Comic History of England* (2 vols, London: Punch Office, 1847–8).

Abir-Am, Pnina. 'How scientists view their heroes: some remarks on the mechanism of myth construction', *Journal of the History of Biology* 15 (1982), 281–315.

Abir-Am, Pnina. 'A historical ethnography of a scientific anniversary in molecular biology: the first protein X-ray photograph (1984, 1934)', *Social Epistemology* 6 (1992), 323–54.

Abrams, M. H. *The Mirror and the Lamp: Romantic Theory and the Critical Tradition* (Oxford: Oxford University Press, 1953).

Acland, Henry W. and John Ruskin. *The Oxford Museum* (London and Orpington: George Allen, 1893).

Addison, Joseph. 'An oration in the defence of the new philosophy', in Bernard de Fontenelle, *A Week's Conversation on the Plurality of Worlds*, transl. William Gardiner (London: for A. Bettesworth and E. Curll, 1737), pp. 197–204.

Adrian, Edgar Douglas. 'Newton's Room in Trinity', *Notes and Records of the Royal Society* 18 (1963), 17–24.

Aikin, John. *A Description of the Country from Thirty to Forty Miles round Manchester* (London: John Stockdale, 1795).

Aiton, E. J. *Leibniz: A Biography* (Bristol and Boston: Adam Hilger, 1985).

Alaya, Flavia. 'Victorian science and the "genius" of woman', *Journal of the History of Ideas* 5 (1977), 261–80.

Albury, W. R. 'Halley's ode on the *Principia* of Newton and the Epicurean revival in England', *Journal for the History of Ideas* 39 (1978), 24–43.

Alexander, H. G. *The Leibniz–Clarke Correspondence* (Manchester: Manchester University Press, 1956).

Algarotti, Francesco. *Sir Isaac Newton's Philosophy Explain'd for the Use of the Ladies*, transl. Elizabeth Carter (2 vols, London, 1739).

Anderson, Benedict. *Imagined Communities: Reflections on the Origin and Spread of Nationalism* (London: Verso, 1983).

Ash, Marinell. 'William Wallace and Robert Bruce: The life and death of a national myth', in Raphael Samuel and Paul Thompson (eds), *The Myths we Live by* (London and New York: Routledge, 1990), pp. 83–94.

Ashworth, William J. '"Labour harder than *thrashing*": John Flamsteed, property and intellectual labour in nineteenth-century England', in Frances Willmoth (ed.), *Flamsteed's Stars: New Perspectives on the Life and Work of the First Astronomer Royal (1646–1719)* (Woodbridge: Boydell Press, 1997).

Atterbury, Francis. *The Epistolary Correspondence, Visitation Charges, Speeches, and Miscellanies* (3 vols, London: J. Nichols, 1783–5).

Ault, Donald D. *Visionary Physics: Blake's Response to Newton* (Chicago and London: University of Chicago Press, 1974).

Avery, Charles. 'Missing, presumed lost: some ivory portraits by David le Marchand', *Country Life* (June 1985), 1562–4.

Avery, Charles. *David Le Marchand 1674–1726: An Ingenious Man for Carving an Ivory* (London: Lund Humphries, 1996).

Babson, Grace. *A Portion of Sir Isaac Newton's Home (1710–1725) Re-erected in the United States* (1939).

Bacon, Francis. *Works*, ed. J. Spedding, R. L. Ellis and D. D. Heath, (7 vols, London: Longmans, 1870–6).

Baillon, Jean-François. 'Aspects de l'impact cultural et idéologique des découvertes de Newton', *Bulletin de la Société d'Études Anglo-Américaines des Dix-septième et Dix-huitième Siècles* 38 (1994), 73–83.

Baker, C. H. Collins. 'Antonio Verrio and Thornhill's early portraiture', *The Connoisseur* 131 (1953), 10–13.

Baker, Malcolm. 'The portrait sculpture', in David McKitterick (ed.), *The Making of the Wren Library* (Cambridge: Cambridge University Press, 1995), pp. 110–37.

Bal, Mieke. 'Telling objects: a narrative perspective on collecting', in John Elsner and Roger Cardinal (eds), *The Cultures of Collecting* (London: Reaktion Books, 1994), pp. 97–115.

Barbauld, Anna Laetitia. *A Legacy for Young Ladies, Consisting of Miscellaneous Pieces, in Prose and Verse* (Boston: David Reed, 1826).

Barlow, Paul. 'Facing the past and present: the National Portrait Gallery and the search for "authentic" portraiture', in Joanna Woodall (ed.), *Portraiture: Facing the Subject* (Manchester and New York: Manchester University Press, 1997), pp. 219–38.

Barthes, Roland. 'The brain of Einstein', in *Mythologies*, transl. Annette Lavers (London: Vintage, 1993), pp. 68–70.

Basbanes, Nicholas A. *A Gentle Madness: Bibliophiles, Bibliomanes, and the Eternal Passion for Books* (New York: Henry Holt, 1995).

Bate, Jonathan. *The Genius of Shakespeare* (London: Picador, 1997).

Bate, Walter Jackson. *John Keats* (London: Hogarth Press, 1992).

Battersby, Christine. *Gender and Genius: Towards a Feminist Aesthetics* (London: The Women's Press, 1989).

Beaven, Derek. *Newton's Niece* (London: Fourth Estate, 1999).

Becker, George. *The Mad Genius Controversy: A Study in the Sociology of Deviance* (Beverly Hills and London: Sage Publications, 1978).

Beddow, Michael. 'Goethe on Genius', in Penelope Murray (ed.), *Genius: The History of an Idea* (Oxford: Basil Blackwell, 1989), pp. 98–112.

Beecher, Jonathan. *Charles Fourier: The Visionary and his World* (Berkeley, Los Angeles and London: University of California Press, 1986).

Bell, Catherine. *Ritual: Perspectives and Dimensions* (New York and Oxford: Oxford University Press, 1997).

Benjamin, Marina. 'Medicine, morality and the politics of Berkeley's tar-water', in A. Cunningham and R. French (eds), *The Medical Enlightenment of the Eighteenth Century*, (Cambridge: Cambridge University Press, 1990), pp. 165–93.

Bensaude-Vincent, Bernadette. 'Between history and memory: centennnial and bicentennial images of Lavoisier', *Isis* 87 (1996), 481–99.

Berkeley, George. *The Works of George Berkeley Bishop of Cloyne*, ed. A. Luce and T. Jessop (London: Thomas Nelson & Sons, 1948–57).

Bernal, J. D. *The Social Function of Science* (London: Routledge, 1939).

Bernal, J. D. *Science in History* (London: Watts & Co., 1954).

Bertoloni Meli, D. 'Caroline, Leibniz, and Clarke', *Journal of the History of Ideas* 60 (1999), 469–86.

Bigg, J. Stanyan. *Night and the Soul* (London: Groombridge & Sons, 1854).

Bindman, David and Malcolm Baker. *Roubiliac and the Eighteenth-century Monument: Sculpture as Theatre* (New Haven and London: Yale University Press, 1995).

Biot, Jean-Baptiste. 'Brewster's *The life of Isaac Newton*', *Journal des Savans* (1832), 199–203, 263–74, 321–39.

Biot, Jean-Baptiste. 'Newton', trans. Howard Elphinstone, in *Lives of Eminent Persons* (London: Baldwin and Cradock, 1833).

Birch, Thomas. *The Heads of Illustrious Persons of Great Britain, On One Hundred and Eight Copper Plates* (London: for John Knapton, 1756).

Blaazer, David. 'Reading the notes: thoughts on the meaning of British paper money', *Humanities Research* 1 (1999), 39–53.

Blake, William. *Blake: Complete Writings with Variant Readings*, ed. Geoffrey Keynes (Oxford and New York: Oxford University Press, 1972).

Blau, Eve. *Ruskinian Gothic: The Architecture of Deane and Woodward 1845–61* (Princeton: Princeton University Press, 1982).

Bond, Donald F. *The Spectator* (5 vols, Oxford: Clarendon Press, 1965).

Bone, Drummond. 'The emptiness of genius: aspects of romanticism', in Penelope Murray (ed.), *Genius: The History of an Idea* (Oxford: Basil Blackwell, 1989), pp. 113–27.

Bongie, Laurence. 'Diderot's *femme savante*', *Studies on Voltaire and the Eighteenth Century* 166 (1977).

Bordes, Philippe and Alain Chevalier, *Catalogue des peintures, sculptures et dessins* (Vizille: Musée de la Révolution française, 1996).

Boskamp, Ulrike. 'Mademoiselle Ferrand méditant sur Newton: von Maurice-Quentin de La Tour zur Rezeption von Newtons *Opticks* in Frankreich vor 1760'. MA thesis, Freie Universität, Berlin (1994).

Boskamp, Ulrike. 'L'arc-en-ciel de Joseph-Marie Vien: oracle d'une théorie de la couleur', in Thomas W. Gaehtgens, Christian Michel, Daniel Rabreau and Martin Schieder (eds), *L'Art et normes sociales au XVIIIe siècle* (Paris: Passages/Passagen, 2001).

Boullée, Étienne-Louis. *L'architecture visionnaire et néoclassique*, ed. J.-M. Pérouse de Montclos (Paris: Hermann, 1993)

Bowles, Geoffrey. 'Physical, human and divine attraction in the life and thought of George Cheyne', *Annals of Science* 31 (1974), 473–88.

Brackenridge, J. Bruce. 'The defective diagram in Newton's *Principia*', in Margaret J. Osler and Paul Lawrence Farber (eds), *Religion, Science and World View: Essays in Honour of Richard S. Westfall* (Cambridge: Cambridge University Press, 1985), pp. 61–93.

Brewer, John. *The Pleasures of the Imagination: English Culture in the Eighteenth Century* (London: HarperCollins, 1997).

Brewster, David. *The Life of Sir Isaac Newton* (London: John Murray, 1831).

Brewster, David. 'Review of Goethe's *Colour Theory*', *Edinburgh Review* 72 (1840), 99–131.

Brewster, David. *Memoirs of the Life, Writings, and Discoveries of Sir Isaac Newton* (2 vols, Edinburgh: Thomas Constable, 1855).

Brian, Denis. *Einstein: A Life* (New York: John Wiley, 1996).

Brown, Peter. *The Cult of the Saints: Its Rise and Function in Latin Christianity* (London: SCM Press, 1981).

Browne, Janet. 'Squibs and snobs: science in humorous British undergraduate magazines around 1830', *History of Science* 30 (1992), 165–97.

Buckley, Theodore Alois. *The Dawnings of Genius Exemplified and Exhibited in the Early Lives of Distinguished Men* (London: Routledge, 1853).

Buds of Genius: Or, Some Account of the Early Lives of Celebrated Characters Who Were Remarkable in their Childhood (London: for Darton, Harvey and Darton, 1816).

Burwick, Frederick. *The Damnation of Newton: Goethe's Colour Theory and Romantic Perception* (Berlin and New York: De Gruyter, 1986).

Butterfield, Herbert. *The Origins of Modern Science 1300–1800* (London: Bell & Sons, 1968).

Byron, George. *Don Juan*, ed. T. G. Steffan, E. Steffan and W. W. Pratt (Harmondsworth: Penguin, 1973).

Cabral, Regis. 'Herbert Butterfield (1900–79) as a Christian historian of science', *Studies in the History and Philosophy of Science* 27 (1996), 547–64.

Cajori, Florian. 'The growth of legend about Sir Isaac Newton', *Science* 59 (1924), 390–2.

Calaprice, Alice. *The Quotable Einstein* (Princeton: Princeton University Press, 1996).

Cambridge, Richard Owen. *Works* (London, 1803).

Cameron, Margaret. *L'influence des* Saisons *de Thomson sur la poésie descriptive en France (1759–1810)* (Geneva: Slatkine Reprints, 1975).

Canning, George. *A Translation of Anti-Lucretius* (London, 1766).

Cantor, Geoffrey. 'Revelation and the cyclical cosmos of John Hutchinson', in Ludmilla Jordanova and Roy Porter (eds), *Images of the Earth: Essays in the History of the Environmental Sciences* (St Giles: British Society for the History of Science, 1978), pp. 3–22.

Cantor, Geoffrey. '*The Analyst* revisited,' *Isis* 75 (1984), 668–83.

Cantor, Geoffrey. 'Weighing light: the role of metaphor in eighteenth-century optical discourse', in Andrew Benjamin, Geoffrey Cantor, and John Christie (eds), *The Figural and the Literal: Problems of Language in the History of Science and Philosophy, 1630–1800* (Manchester: Manchester University Press, 1989), pp. 124–46.

Cantor, Geoffrey. 'Where the statue stood: celebrations of Faraday, 1867–1931', unpublished paper presented at the Royal Institution conference on *Science and its Publics*, 22 September 1994.

Cantor, Geoffrey. 'The scientist as hero: public images of Michael Faraday', in Michael Shortland and Richard Yeo (eds), *Telling Lives in Science: Essays on Scientific Biography* (Cambridge: Cambridge University Press, 1996), pp. 171–93.

Cantor, Geoffrey and Michael Hodge. 'Introduction', in Geoffrey Cantor and Michael Hodge (eds), *Conceptions of Ether: Studies in the History of Ether Theories, 1740–1900*, (Cambridge: Cambridge University Press, 1981), pp. 1–60.

Carlyle, Thomas. 'Signs of the times', *Edinburgh Review* 49 (1829), 439–59.

Cassirer, Ernst. *The Philosophy of the Enlightenment*, transl. Fritz C. A. Koelln and James P. Pettegrove (Boston: Beacon Press, 1951).

Casteras, Susan P. 'Excluding women: the cult of the male genius in Victorian painting', in Linda M. Shires (ed.), *Rewriting the Victorians: Theory, History, and the Politics of Gender* (London and New York: Routledge, 1992), pp. 116–46.

Chambers, Robert. *The Book of Days* (London and Edinburgh: W. & R. Chambers; 2 vols, 1864).

Chandler, James. *England in 1819: the Politics of Literary Culture and the Case of Romantic Historicism* (Chicago and London: University of Chicago Press, 1998).

Charlton, D. G. *New Images of the Natural in France: A Study in European Cultural History 1750–1800* (Cambridge: Cambridge University Press, 1984).

Chartres, John. 'No English Calvados? English distillers and the cider industry in the seventeenth and eighteenth centuries', in John Chartres and David Hey (eds), *English Rural Society, 1500–1800: Essays in Honour of Joan Thirsk* (Cambridge: Cambridge University Press, 1990), pp. 313–42.

Chênedollé, Charles. *Le Génie de l'homme* (Paris: Charles Gosselin, 1822).

Chénier, André. *Oeuvres complètes*, ed. Gérard Walter (Paris: Gallimard, 1958).

Chénier, Marie-Josèphe. *Oeuvres* (8 vols, Paris: Guillaume, 1824–6).

Cheyne, George. *The English Malady*, ed. Roy Porter (London and New York: Tavistock/Routledge, 1991).

Christie, John R. R. 'Sir David Brewster as a historian of science', in A. D. Morrision-Low and J. R. R. Christie (eds), *'Martyr of science': Sir David Brewster 1781–1868* (Edinburgh: Royal Scottish Museum Studies, 1984), pp. 52–6.

Clubbe, John. *Physiognomy: Being a Sketch only of a Larger Work upon the Same Plan: Wherein the Different Tempers, Passions, and Manners of Men, will be Particularly Considered* (London: for R. & J. Dodsley, 1763).

Cobbett, William (ed.), *The Parliamentary History of England from the Earliest Period to the Year 1803*, vols 13–36 (London: Longman & Co., 1812–20).

Cohen, I. Bernard. 'Commemorations and memorials: Isaac Newton 1727–1977', *Vistas in Astronomy* 22 (1979), 381–94.

Cohen, I. Bernard. *Science and the Founding Fathers: Science in the Political Thought of Jefferson, Franklin, Adams and Madison* (New York and London: W. W. Norton, 1995).

Colton, Judith. 'Kent's Hermitage for Queen Caroline at Richmond', *Architectura* (1974), 181–91.

Combe, Andrew. *The Physiology of Digestion Considered with Relation to the Principles of Dietetics* (Edinburgh: Maclachlan, Stewart & Co., 1845).

Condillac, Étienne Bonnot. *An Essay on the Origin of Human Knowledge*, transl. Thomas Nugent (London: J. Nourse, 1756).

Condillac, Étienne Bonnot. *Philosophical Writings*, transl. Franklin Philip and Harlan Lane (Hillsdale, NJ, and London: Lawrence Erlbaum, 1982).

Condorcet, Antoine-Nicolas de. *Sketch for a Historical Picture of the Progress of the Human Mind*, transl. June Barraclough (London: Weidenfeld and Nicolson, 1955)

Conisbee, Paul. *Painting in Eighteenth-century France* (Oxford: Phaidon, 1981).

Connerton, Paul. *How Societies Remember* (Cambridge: Cambridge University Press, 1989).

Cooper, Thomas. *The Purgatory of Suicides* (London: Jeremiah How, 1845).

Cooper, Thomas. *Cooper's Journal: Or, Unfettered Thinker and Plain Speaker for Truth, Freedom, and Progress* (London: James Watson, 1850).

Cooper, Thomas. *The Triumphs of Perseverance and Enterprise: Recorded as Examples for the Young* (London: Darton & Co., 1856).

Cooper, Thomas. *The Life of Thomas Cooper* (London: Hodder & Stoughton, 1873).

Cooper, Thomas. *The Paradise of Martyrs: A Faith Rhyme* (London: Hodder & Stoughton, 1873).

Cooter, Roger. *The Cultural Meaning of Popular Science: Phrenology and the Organization of Consent in Nineteenth-century Britain* (Cambridge: Cambridge University Press, 1984).

Craik, G. L. *The Pursuit of Knowledge under Difficulties, Illustrated by Anecdotes* (2 vols, London, Charles Knight, 1830–1).

Croly, George. 'The British fleet', *Blackwood's Edinburgh Magazine* 55 (1844), 462–82.

Crompton, Samuel. 'On the portraits of Sir Isaac Newton; and particularly on one of him by Kneller, painted about the time of publication of the *Principia*, and representing him as he was in the prime of life', *Proceedings of the Literary and Philosophical Society of Manchester* 6 (1866), 1–7.

Crompton, Samuel. [untitled]. *Proceedings of the Literary and Philosophical Society of Manchester* 7 (1867), 3–6.

Crow, Thomas. *Painters and Public Life in Eighteenth-century Paris* (New Haven and London: Yale University Press, 1985).

Cunningham, Andrew. 'How the *Principia* got its Name; or, taking Natural Philosophy seriously', *History of Science* 29 (1991), 377–92.

Curtis, Penelope, Peter Funnell and Nicola Kalinsky. *Return to Life: A New Look at the Portrait Bust* (Leeds: Henry Moore Institute, 2001).

Dainville, François de. 'L'Enseignement scientifique dans les collèges Jésuites', in René Taton (ed.), *Enseignement et diffusion des sciences en France au XVIII*[e] *siècle* (Paris: Hermann, 1964), pp. 27–65.

D'Alembert, Jean le Rond. *Preliminary Discourse to the Encyclopedia of Diderot*, transl. Richard N. Schwab and Walter E. Rex (Indianapolis and New York: Bobbs-Merrill, 1963).

Daniels, Stephen. *Joseph Wright* (London: Tate Gallery, 1999).

Darnton, Robert. *The Forbidden Best-sellers of Pre-Revolutionary France* (London: Fontana, 1997).

Dawson, Aileen. *Masterpieces of Wedgwood in the British Museum* (London: British Museum Publications, 1984).

De Morgan, Augustus. 'Theory of probabilities – part II', *Dublin Review* 3 (1837), 237–48.

De Morgan, Augustus. *Essays on the Life and Work of Newton*, ed. Philip E. B. Jourdain (Chicago and London: Open Court, 1914).

De Morgan, Augustus. *A Budget of Paradoxes* (2 vols, London: Open Court Company, 1915).

De Nardis, Luigi. *Saint-Lambert: Scienza e paesaggio nella poesia del Settecento* (Rome: Edizione dell'Ateneo, 1961).

De Nora, Tia and Hugh Mehan. 'Genius: a social construction, the case of Beethoven's initial recognition', in Theodore R. Sarbin and John I. Kitsuse (eds), *Constructing the Social* (London: Thousand Oaks and New Delhi: Sage, 1994), pp. 157–73.

Debrie, Christine. *Maurice Quentin de la Tour: 'Peintre de portraits en pastel' 1704–1788 au Museé Antonie Lécuyer de Saint Quentin* (Saint-Quentin: Albaron, 1991).

Delisle de Sales, Jean. *De la philosophie de la nature, ou traité de morale pour l'espèce humaine* (10 vols, Paris, 1804).

Dellheim, Charles. *The Face of the Past: The Preservation of the Medieval Inheritance in Victorian England* (Cambridge: Cambridge University Press, 1982).

Dening, Greg. *The Death of William Gooch: A History's Anthropology* (Melbourne: Melbourne University Press, 1995).

Desaguliers, John Theophilus. *The Newtonian System of the World, the Best Model of Government: An Allegorical Poem* (London, 1728).

Desaguliers, John Theophilus. *A Course of Experimental Philosophy* (2 vols, London, 1744–5).

Dhombres, Jean. 'Culture scientifique et poésie aux alentours de la Révolution française', in Claude Blanckaert, Jean-Louis Fischer and Roselyne Rey (eds), *Nature, histoire, société: essais en hommage à Jacques Roger* (Paris: Klincksieck, 1995), pp. 341–67.

Dhombres, Nicole and Jean Dhombres. *Naissance d'un pouvoir: Sciences et savants en France (1793–1824)* (Paris: Éditions Payot, 1989).

Diderot, Denis. *Oeuvres complètes* (Paris: Hermann, 1975–89).

Diderot, Denis. *The Indiscreet Jewels*, transl. Sophie Hawkes (New York: Marsilio, 1993).

Dieckmann, Herbert. 'Diderot's conception of genius', *Journal for the History of Ideas* 2 (1941), 151–82.

Dimoff, Paul. *La vie et l'œuvre d'André Chénier jusqu'à la Révolution française 1762–1790* (2 vols, Paris: E. Droz, 1936).

Disraeli, Isaac. *Literary Character of Men of Genius* (London: Frederick Warne & Co., 1881).

Ditchburn, R. W. 'Newton's illness of 1692–3', *Notes and Records of the Royal Society* 35 (1980), 1–16.

Ditton, Humphry. *The General Laws of Nature and Motion; with their Application to Mechanicks* (London, 1705).

Dobbs, Betty Jo Teeter. *The Janus Faces of Genius: The Role of Alchemy in Newton's Thought* (Cambridge: Cambridge University Press, 1991).

Dobson, Michael. *The Making of the National Poet: Shakespeare, Adaptation, and Authorship, 1660–1769* (Oxford: Clarendon Press, 1992).

D'Oench, Ellen G. *The Conversation Piece: Arthur Devis and His Contemporaries* (New Haven: Yale Center for British Art, 1980).

Dollond, John. 'A letter to James Short, A.M. F.R.S. concerning a mistake in M. Euler's theorem for correcting the aberrations in the object-glasses of refracting telescopes', *Philosophical Transactions* 48 (1753), 289–91.

Dubos, René and Jean Dubos. *The White Plague: Tuberculosis, Man and Society* (London: Victor Gollancz, 1953).

Du Châtelet, Émilie. *Principes mathématiques de la Philosophie Naturelle* (2 vols, Paris: Éditions Jacques Gabay, 1990 facsimile of 1759 edition).

Dundon, Stanislaus John Sherman. 'Philosophical resistance to Newtonianism on the Continent, 1700–1760', PhD dissertation, St John's University, 1972.

Dunning-Davies, Jeremy. 'Popular status and scientific influence: another angle on "the Hawking phenomenon"', *Public Understanding of Science* 2 (1993), 85–6.

Duportal, Jeanne. *Bernard Picart 1673 à 1733* (Paris and Brussels: Éditions g. Van Oest, 1928).

Earman, John and Clark Glymour, 'Relativity and eclipses: the British eclipse expeditions of 1919 and their predecessors', *Historical Studies in Physical Science* 11 (1980), 49–85.

Edgar, J. G. *The Boyhood of Great Men Intended as an Example to Youth* (London: David Bogue, 1853).

Edgeworth, Richard Lovell and Maria Edgeworth. *Essay on Irish Bulls* (London, 1802).

Edleston, Joseph. *Correspondence of Sir Isaac Newton and Professor Cotes, Including Letters of Other Eminent Men, Now First Published from the Originals in the Library of Trinity College, Cambridge* (Cambridge: J. W. Parker, 1850).

Ehrard, Jean. *L'idée de nature en France dans la première moitié du XVIIIe siècle* (Paris: SEVPEN, 1963).

Einstein, Albert. *Out of my Later Years* (New York: Philosophical Library, 1950).

Ellis, Florence. *Character Forming in School* (London: Longman, Green & Co., 1907).

Ellison, Henry. *Stones from the Quarry; Or, Moods of Mind* (London: Provost & Co., 1875).

Emerson, William. *The Principles of Mechanics* (London: J. Richardson, 1758).

Essick, Robert N. 'Blake's Newton', *Blake Studies* 3(2), 149–62.

Etlin, Richard A. *The Architecture of Death: The Transformation of the Cemetery in Eighteenth-Century Paris* (Cambridge, MA, and London: MIT Press, 1984).

Euler, Leonhard. *Letters to a German Princess, on Different Subjects in Physics and Philosophy* (2 vols, London, 1795).

Evans, James. 'Fraud and illusion in the anti-Newtonian rear guard', *Isis* 87 (1996), 74–107.

Evelyn, John. *Sylva, or a Discourse of Forest-trees, and the Propagation of Timber in His Majesties Dominions* (London, 1664).

Falkner, Alexander. 'Introduction to the study of phrenology', *Phrenological Almanac* 1 (1842), 1–16.

Fara, Patricia. *Sympathetic Attractions: Magnetic Practices, Symbolism, and Beliefs in Eighteenth-century England* (Princeton: Princeton University Press, 1996).

Farrer, J. A. *Literary Forgeries* (New York: Longman, Green & Co., 1907).

Fauvel, John *et al.*, *Let Newton Be! A New Perspective on His Life and Works* (Oxford: Oxford University Press, 1988).

Fontanes, Louis. 'Essai sur l'astronomie', in *Oeuvres* (2 vols, Paris: La Hachette, 1859).

Force, James E. *William Whiston: Honest Newtonian* (Cambridge: Cambridge University Press, 1985).

Forgan, Sophie. 'The architecture of science and the idea of a university', *Studies in the History and Philosophy of Science* 20 (1989), 405–34.

Fortune, Brandon Brame and Deborah J. Warner. *Franklin and His Friends: Portraying the Man of Science in Eighteenth-century America* (Philadelphia: University of Pennsylvania Press, 1999).

Freudenthal, Gideon. *Atom and Individual in the Age of Newton: On the Genesis of the Mechanistic World View* (Dordrecht: D. Reidel, 1986).

Fried, Michael. *Absorption and Theatricality: Painting and Beholder in the Age of Diderot* (Berkeley, Los Angeles and London: University of California Press, 1980).

Friedman, Alan J. and Carol C. Donley. *Einstein as Myth and Muse* (Cambridge: Cambridge University Press, 1985).

Froom, Leroy E. *The Prophetic Faith of our Fathers* (4 vols, Washington, DC: Review and Herald, 1950–4).

Fumaroli, Marc. 'Le génie de la langue française', in Pierre Nora (ed.), *Les Lieux de mémoire* (3 vols, Paris: Gallimard, 1997), vol. 3, pp. 4623–85.

Furst, Lilian R. *Romanticism in Perspective: A Comparative Study of Aspects of the Romantic Movements in England, France and Germany* (London, Melbourne and Toronto: Macmillan, 1969).

Gage, John. *Colour and Culture: Practice and Meaning from Antiquity to Abstraction* (London: Thames and Hudson, 1993).

Gage, John. *Colour and Meaning: Art, Science and Symbolism* (London: Thames and Hudson, 1999).

Galloway, T. 'French and English biographies of Newton', *Foreign Quarterly Review* 12 (1833), 1–27.

Gardner, Martin. *Fads and Fallacies in the Name of Science* (New York: Dover, 1957).

Gascoigne, John. 'From Bentley to the Victorians: the rise and fall of British Newtonian natural theology', *Science in Context* 2 (1988), 219–56.

Gaskell, Elizabeth. *Mary Barton* (Harmondsworth: Penguin, 1970).

Gaze, John. *Figures in a Landscape: A History of the National Trust* (London: Barrie and Jenkins in association with the National Trust, 1988).

Geary, Patrick. 'Sacred commodities: the circulation of medieval relics', in Arjun Appadurai (ed.), *The Social Life of Things: Commodities in Cultural Perspective* (Cambridge: Cambridge University Press, 1986), pp. 169–91.

Gebelin, François and André Morize. *Correspondance de Montesquieu* (2 vols, Paris: Champion, 1913–14).

Genlis, Stéphanie. *Adelaide and Theodore; Or, Letters on Education* (3 vols, Dublin, 1783).

George, M. Dorothy. *Catalogue of Political and Personal Satires Preserved in the Department of Prints and Drawings in the British Museum* (London: Trustees of the British Museum, 1935–54).

Gerard, Alexander. *An Essay on Genius* (London: for W. Strahan, 1774).

Gillispie, Charles Coulston. *The Edge of Objectivity: An Essay in the History of Scientific Ideas* (Princeton: Princeton University Press, 1960).

Gilpin, William. *A Dialogue upon the Gardens of the Right Honourable the Lord Viscount Cobham, at Stow in Buckinghamshire* (London: for B. Seeley, 1748).

Girardin, Marquis de. *Promenade ou itinéraire des Jardins d'Ermenonville* (Paris: Mérigot, 1788).

Gjertsen, Derek. *The Newton Handbook* (London and New York: Routledge and Kegan Paul, 1986).

Goodman, Dena. *The Republic of Letters: A Cultural History of the French Enlightenment* (Ithaca and London: Cornell University Press, 1994).

Gordon, Margaret Maria. *The Home Life of Sir David Brewster* (Edinburgh, 1869).

Gray, John M. *James and William Tassie: A Biographical and Critical Sketch with a Catalogue of their Portrait Medallions of Modern Personages* (Edinburgh: Walter Greencock Patterson, 1894).

Greenberg, John L. *The Problem of the Earth's Shape from Newton to Clairaut: The Rise of Mathematical Science in Eighteenth-century Paris and the Fall of 'Normal' Science* (Cambridge: Cambridge University Press, 1995).

Greenberg, Mark L. 'Blake's "Science"', *Studies in Eighteenth-century Culture* 12 (1983), 115–30.

Grieder, Josephine. *Anglomania in France 1740–1789: Fact, Fiction, and Political Discourse* (Geneva: Droz, 1985).

Gudin de la Brenellerie, Paul-Philippe. *L'Astronomie, poëme en quatre chants* (Paris: Firmin Didot, 1810).

Guerlac, Henry. *Newton on the Continent* (Ithaca and London: Cornell University Press, 1981).

Guerrini, Antonia. 'The Tory Newtonians: Gregory, Pitcairne and their circle,' *Journal of British Studies* 25 (1986), 288–311.

Guest, Harriet. *A Form of Sound Words* (Oxford: Clarendon Press, 1989).

Guicciardini, Niccolò. *Reading the* Principia: *The Debate on Newton's Mathematical Methods for Natural Philosophy from 1687 to 1736* (Cambridge: Cambridge University Press, 1999).

Hagner, Michael. 'Kluge Köpfe und geniale Gehirne: zur Anthrolopolgie des Wissenschaftlers im 19. Jahrhundert', in Hans Erich Bödeker, Peter Hanns Reill and Jürgen Schlumbohm (eds), *Wissenschaft als kulturelle Praxis, 1750–1900* (Göttingen: Vandenhoek & Ruprecht, 1999), pp. 299–33.

Hales, Stephen. *Vegetable Staticks*, ed. M A Hoskin (London: Oldbourne, 1969).

Halévy, Elie. *The Growth of Philosophical Radicalism*, transl. Mary Morris (London: Faber and Faber, 1972).

Halford, Henry. 'The deaths of some eminent philosophers', *Wesleyan Methodist Magazine* 15 (3rd series) (1836), 277–82.

Halfpenny, Pat. *English Earthenware Figures 1740–1840* (Woodbridge: Antique Collectors Club, 1991).

Hall, A. Rupert. *Ballistics in the Seventeenth Century: A Study in the Relations of Science and War with Reference Particularly to England* (Cambridge: Cambridge University Press, 1952).

Hall, A. Rupert. 'Newton in France: a new view', *History of Science* 13 (1975), 233–50.

Hall, A. Rupert. *Philosophers at War: The Quarrel between Newton and Leibniz* (Cambridge: Cambridge University Press, 1980).

Hall, A. Rupert. *Isaac Newton: Adventurer in Thought* (Oxford and Cambridge, MA: Blackwell, 1992).

Hall, A. Rupert. *Isaac Newton: Eighteenth-century Perspectives* (Oxford, New York and Tokyo: Oxford University Press, 1999).

Hall, Thomas S. 'On biological analogs of Newtonian paradigms', *Philosophy of Science* 35 (1968), 6–27.

Harrison, Frederic (ed.). *The New Calendar of Great Men* (London: Macmillan, 1892).

Hartill, Isaac. *Recollections of Isaac Newton* (London: James Clarke & Co., 1914).

Haskell, Francis. 'The apotheosis of Newton in art', in F. Haskell, *Past and Present in Art and Taste: Selected Essays* (New Haven and London: Yale University Press, 1987), pp. 1–15.

Hawking, Stephen W. 'Newton's *Principia*', in S. W. Hawking and W.

Israel (eds), *Three Hundred Years of Gravitation* (Cambridge: Cambridge University Press, 1987), pp. 1–4.

Hawking, Stephen W. *A Brief History of Time* (London: Bantam Press, 1988).

Hawkins, Edward, Augustus W. Franks and Herbert A Grueber, *Medallic Illustrations of the History of Great Britain and Ireland to the Death of George II* (2 vols, London: British Museum, 1885).

Haydon, Benjamin Robert. *Lectures on Painting and Design* (London: Longman, Brown, Green and Longman, 1844).

Haydon, Benjamin Robert. *The Diary of Benjamin Robert Haydon*, ed. Willard Bissell Pope (5 vols, Cambridge, MA: Harvard University Press, 1960–3).

Hayley, William. *The Life of George Romney, Esq.* (Chichester: T. Payne, 1809).

Hazlitt, William. *The Complete Works of William Hazlitt*, ed. P. P. Howe (21 vols, London: J. M. Dent, 1930–4).

Hazlitt, William. *Selected Writings*, ed. Jon Cook (New York and Oxford: Oxford University Press, 1983).

Healy, Chris. *From the Ruins of Colonialism: History as Social Memory* (Cambridge: Cambridge University Press, 1997).

Heath-Stubbs, John and Phillips Salman. *Poems of Science* (Harmondsworth: Penguin, 1984).

Heffernan, James A. W. 'The English Romantic perception of colour', in Karl Kroeber and William Walling (eds), *Images of Romanticism: Verbal and Visual Affinities* (New Haven and London: Yale University Press, 1978), pp. 133–48.

Henig, Martin, Diana Scarisbrick and Mary Whiting. *Classical Gems, Ancient and Modern Intaglios and Cameos* (Cambridge: Cambridge University Press, 1994).

Herschel, Mary Cornwallis. *Memoir and Correspondence of Caroline Herschel* (London, 1876).

Hesse, Carla. 'Enlightenment epistemology and the laws of authorship in Revolutionary France, 1777–1793', *Representations* 30 (1990), 109–37.

Hessen, Boris. 'The social and economic roots of Newton's "Principia"', in *Science at the Crossroads* (London: Kniga, 1931).

Hinks, Roger. 'A symbolic portrait of Descartes', *Journal of the Warburg and Courtauld Institutes* 3 (1939–40), 156.

Hitchcock, Edward. 'Blessings of temperance in food: a sermon', from

American National Preacher, reproduced in *Wesleyan Methodist Magazine* 15 (3rd series) (1836), 660–73.

Hodgskin, Thomas. *Four Lectures delivered at the London Mechanics' Institution* (London: for Charles and William Tait, 1827).

Hogg, Robert. *The Herefordshire Pomona, Containing Coloured Figures and Descriptions of the Most Esteemed Kinds of Apples and Pears* (2 vols, Hereford: Jakeman and Carver, 1876–85).

Holderness, Graham. 'Bardolatry: or, the cultural materialist's guide to Stratford-upon-Avon', in Graham Holderness (ed.), *The Shakespeare Myth* (Manchester: Manchester University Press, 1988), pp. 2–15.

Holmes, Richard. *Coleridge: Darker Reflections* (London: Flamingo, 1999).

Home, Henry. *Sketches of the History of Man* (4 vols, Edinburgh, 1788).

Horne, Richard Hengist. *Historical Tragedy: And Other Poems* (London: George Rivers, 1875).

Hughes, G. Bernard. 'Notable earthenware figures', *Country Life Annual* (1957), pp. 52–61.

Hughes, G. Bernard. 'Portrait busts in black basaltes', *Country Life* 132 (1962), 360–1.

Hume, David. 'A dissertation on the passions', in *The Philosophical Works*, ed. Thomas Hill Green and Thomas Hodge Grose (4 vols, Darmstadt: Scientia Verlag Aalen, 1992, reprint of 1882 London edition), pp. 137–66.

Hunt, John Dixon. 'Emblem and expressionism in the eighteenth-century landscape garden', *Eighteenth-century Studies* 4 (1971), 294–317.

Hunt, John Dixon. *The Figure in the Landscape: Poetry, Painting, and Gardening during the Eighteenth Century* (Baltimore and London: Johns Hopkins University Press, 1976).

Hunt, John Dixon and Peter Willis (eds). *The Genius of the Place: The English Landscape Garden 1620–1820* (London: Paul Elek, 1975).

Hunter, Michael. 'Robert Boyle and the dilemma of biography in the age of the Scientific Revolution', in Michael Shortland and Richard Yeo (eds), *Telling Lives in Science: Essays on Scientific Biography* (Cambridge: Cambridge University Press, 1996), pp. 115–37.

Hutchinson, John. *Moses's Principia* (London, 1724–7).

Iliffe, Rob. '"Aplatisseur du monde et de Cassini": Maupertuis, precision measurement, and the shape of the earth in the 1730s', *History of Science* 31 (1993), 335–75.

Iliffe, Rob. '"Is he like other men?" The meaning of the *Principia mathematica*, and the author as idol', in Gerald Maclean (ed.), *Culture and Society in the Stuart Restoration* (Cambridge: Cambridge University Press, 1995), pp. 159–76.

Iliffe, Rob. 'A "connected system"? The snare of a beautiful hand and the unity of Newton's archive', in Michael Hunter (ed.), *Archives of the Scientific Revolution: The Formation and Exchange of Ideas in Seventeenth-century Europe* (Woodbridge: Boydell Press, 1998), pp. 137–57.

Iliffe, Rob. 'Isaac Newton: Lucatello Professor of Mathematics', in Christopher Lawrence and Steven Shapin (eds), *Science Incarnate: Historical Embodiments of Natural Knowledge* (Chicago and London: University of Chicago Press, 1998), pp. 121–55.

Inwards, Jabez. *Phrenological Annotations* (London: William Tweedie, 1864).

Jaffe, Kineret J. 'The concept of genius: its changing role in eighteenth-century French aesthetics', *Journal of the History of Ideas* 41 (1980), 579–99.

Jalland, Pat. *Death in the Victorian Family* (Oxford: Oxford University Press, 1996).

Jay, Martin. *Downcast Eyes: The Denigration of Vision in Twentieth-century French Thought* (Berkeley and London: University of California Press, 1993).

Jeans, James. 'Isaac Newton', *Nature* (supplement for 26 March) 119 (1927), 28–30.

Jenkyns, Richard. *Dignity and Decadence: Victorian Art and the Classical Inheritance* (London: Fontana Press, 1992).

Jerman, William. 'Nonsense! A miscellany about love', *Bentley's Miscellany* 4 (1838), 167–73.

Johns, Adrian. *The Nature of the Book: Print and Knowledge in the Making* (Chicago and London: University of Chicago Press, 1998).

Jones, W. Powell. *The Rhetoric of Science: A Study of Scientific Ideas and Imagery in Eighteenth-century English Poetry* (London: Routledge & Kegan Paul, 1966).

Jopling, Joseph. 'Newton's home in the year 1727', *Leisure Hour* 11 (1862), 584–7.

Jordanova, Ludmilla. 'Science and nationhood: cultures of imagined communities', in Geoffrey Cubitt (ed.), *Imagining Nations* (Manchester and New York: Manchester University Press, 1998), pp. 192–211.

Jordanova, Ludmilla. *Defining Features: Scientific and Medical Portraits 1660–2000* (London: Reaktion Books, 2000).

Kant, Immanuel. *Anthropology from a Pragmatic Point of View*, transl. Victor Lyle Dowdell (Carbondale and Edwardsville: South Illinois University Press, 1978).

Kant, Immanuel. *Critique of Judgement*, transl. Werner S. Pluhar (Indianapolis: Hackett Publishing Company, 1987).

Kant, Immanuel. 'An answer to the question: What is enlightenment?' in *Practical Philosophy*, ed. Mary J Gregor (Cambridge: Cambridge University Press, 1996), pp. 13–22.

Keesing, R. G. 'The history of Newton's apple tree', *Contemporary Physics* 39 (1998), 377–91.

Kemble, John. *State Papers and Correspondence Illustrative of the Social and Political State of Europe from the Revolution to the Accession of the House of Hanover* (London: John J. Parker, 1857).

Kerslake, John. 'Sculptor and patron? Two portraits by Highmore', *Apollo* 95 (January 1972), 25–9.

Keyes, J. Gregory. *Newton's Cannon* (New York: Ballantine, 1998).

Keynes, John Maynard. 'Newton, the man', in *The Royal Society Newton Tercentenary Celebrations* (Cambridge: Cambridge University Press, 1947), pp. 27–34.

King, Edmund Fillingham. *A Biographical Sketch of Isaac Newton* (Grantham, 1858).

King, Edmund Fillingham. *Ten Thousand Wonderful Things* (London: Ward and Lock, 1859).

Kinns, Samuel. *'Six hundred Years'; Or, Historical Sketches of Eminent Men and Women Who Have More or Less Come into Contact with the Abbey and Church of Holy Trinity, Minories, from 1293 to 1893* (London: Cassell & Co., 1898).

Kistler, Mark O. *Drama of the Storm and Stress* (New York: Twayne Publishers, 1969).

Klein, Jürgen. 'Genius, ingenium, imagination: aesthetic theories of production from the Renaissance to Romanticism', in Frederick Burwick and Jürgen Klein (eds), *The Romantic Imagination: Literature and Art in England and Germany*, (Amsterdam and Atlanta: Rodopi, 1996), pp. 19–62.

Klopp, Onno. *Die Werke von Leibniz* (11 vols, Hanover: Klindworth's Verlag, 1864–84).

Krätz, Otto. *Goethe und die Naturwissenschaften* (Munich: Callwey, 1992).

Kretschmer, Ernst. *Körperbau und Charakter: Untersuchungen zum Konstitutionsproblem und zur Lehre von den Temperamenten* (Berlin: Julius Springer, 1921).

Kris, Ernst and Otto Kurz. *Legend, Myth, and Magic in the Image of the Artist: A Historical Experiment* (New Haven and London: Yale University Press, 1979).

Kriz, Kay Dian. *The Idea of the English Landscape Painter: Genius as Alibi in the Early Nineteenth Century* (New Haven and London: Yale University Press, 1997).

Laudan, Larry L. 'Thomas Reid and the Newtonian turn of British methodological thought', in Robert E. Butts and John W. Davis (eds), *The Methodological Heritage of Newton* (Oxford: Basil Blackwell, 1970), pp. 103–31.

Le Dantec, Denise and Jean-Pierre le Dantec. *Reading the French Garden: Story and History* (Cambridge, MA, and London: MIT Press, 1987).

Leighton, Robert. *Records and Other Poems* (London: C. Kegan Paul & Co., 1880).

Lemercier, Népomucène Louis. *L'Atlantiade, ou la théogonie Newtonienne, poëme en six chants* (Paris: Pichard, 1812).

Leppert, Richard. *Music and Image: Domesticity, Ideology and Socio-cultural Formation in Eighteenth-century England* (Cambridge: Cambridge University Press, 1988).

Lightman, Bernard. 'The voices of nature: popularizing Victorian science', in Bernard Lightman (ed.), *Victorian science in Context* (Chicago and London: University of Chicago Press, 1997), pp. 187–211.

Linton, William James. *Catoninetales: a domestic epic: by Hattie Brown: a young lady of colour lately deceased at the age of 14* (Hamden, CT: Appledore US, 1891).

Lives of Illustrious Men (London: Thomas Nelson, 1851).

Lives of Learned and Eminent Men, Taken from Authentic Sources, Adapted to the Use of Children of Four Years Old and Upwards (1821).

Llewellyn, Roger. *The Art of Death: Visual Culture in the English Death Ritual c.1500–c.1800* (London: Reaktion Books, 1991).

Lloyd-Jones, I. D. 'Charles Fourier: faithful pupil of the Enlightenment', in Peter Gilmour (ed.), *Philosophers of the Enlightenment* (Edinburgh: Edinburgh University Press, 1989), pp. 151–78.

Lowenthal, David. *The Past is a Foreign Country* (Cambridge: Cambridge University Press, 1985).

Lucretius. *His Six Books of Epicurean Philosophy*, transl. Thomas Creech (London, 1700).

Macaulay, Catherine. *A Modest Plea for the Property of Copyright* (London: Edward and Charles Dilly, 1774) reproduced in *The Literary Property Debate: Eight Tracts, 1774–1775* (New York and London: Garland, 1974).

Maclaurin, Colin. *An Account of Sir Isaac Newton's Philosophical Discoveries, in Four Books* (London, 1748).

MacLeod, Christine. 'The paradoxes of patenting: invention and its diffusion in 18th- and 19th-century Britain, France, and North America', *Technology and Culture* 32 (1991), 885–910.

MacLeod, Christine. 'Concepts of invention and the patent controversy in Victorian Britain', in Robert Fox (ed.), *Technological Change: Methods and Themes in the History of Technology* (Amsterdam: Harwood Academic, 1996), pp. 137–53.

MacLeod, Christine. 'James Watt, heroic invention and the idea of the industrial revolution', in Maxine Berg and Kristine Bruland (eds), *Technological Revolutions in Europe: Historical Perspectives* (Cheltenham and Northampton, MA: Edward Elgar, 1998), pp. 96–116.

MacLeod, Roy. M. 'Whigs and savants: reflections on the reform movement in the Royal Society, 1830–48', in Ian Inkster and Jack Morrell (eds), *Metropolis and Province: Science in British Culture, 1780–1850* (London: Hutchinson, 1983), pp. 55–90.

MacPike, Eugene Fairfield. *Correspondence and Papers of Edmond Halley* (London: Taylor and Francis, 1937).

Mais, S. P. B. 'Isaac Newton at Woolsthorpe'. *Country Life* 92 (1942), 1224–5.

Majewski, Henry F. *The Preromantic Imagination of L.-S. Mercier* (New York: Humanities Press, 1971).

Malandian, Pierre. *Delisle des Sales: philosophe de la nature (1741–1816)* (2 vols, Oxford: Voltaire Foundation, 1982)

Mali, Joseph. 'Narrative, myth and history', *Science in Context* 7 (1994), 121–42.

Mallet, David. 'The excursion', in Robert Anderson (ed.), *A Complete Edition of the Poets of Great Britain* (13 vols, London, 1795), vol. 9, pp. 687–95.

Mallet, J. V. G. 'Some portrait medallions by Roubiliac', *Burlington Magazine* 104 (1962), 153–8.

Malthus, Thomas R. *Principles of Political Economy Considered with a View to Their Practical Application* (London: John Murray, 1820).

Mann, Elizabeth L. 'The problem of originality in English literary criticism, 1750–1800', *Philological Quarterly* 18 (1939), 97–118.

Manuel, Frank E. *The New World of Henri Saint-Simon* (Cambridge, MA: Harvard University Press, 1956).

Manuel, Frank E. *Isaac Newton: Historian* (Cambridge, MA: Belknap Press, 1963).

Manuel, Frank E. *Portrait of Isaac Newton* (Cambridge, MA: Harvard University Press, 1968).

Manuel, Frank E. and Fritzie P. Manuel. *Utopian Thought in the Western World* (Oxford: Basil Blackwell, 1979).

Markley, Robert. *Fallen Languages: Crises of Representation in Newtonian England, 1660–1740* (Ithaca and London: Cornell University Press, 1993).

Martin, Benjamin. *Biographia Philosophica* (London, 1764).

Martin, Louis Aimé. *Lettres à Sophie* (Paris: Charpentier, 1842).

Massey, Gerald. *Craigcrook Castle* (London: David Bogue, 1856).

Masters, Walter E. *The Huguenot Church and Sir Isaac Newton's House, being the History of Orange Street Chapel from 1688 to 1775, and of Sir Isaac Newton's House adjoining.* 1910.

Matthaei, Ruprecht. *Corpus der Goethezeichnungen, Band 5A* (Leipzig: E. A. Seemann, 1963).

Maty, Matthew. *Authentic Memoirs of the Life of Richard Mead MD* (London: for J. Whiston and B. White, 1755).

Maude, Thomas. *Wensley-dale; or, Rural Contemplations: A Poem* (London: 1780).

Maude, Thomas. *Viator, a Poem: or, A Journey from London to Scarborough, By the Way of York* (London: B. White *et al.*, 1782).

McKendrick, Neil, John Brewer and J. H. Plumb, *The Birth of a Consumer Society: the Commercialization of Eighteenth-century England* (London: Europa, 1982).

McKie, D. and G. R. de Beer (a), 'Newton's apple', *Notes and Records of the Royal Society* 9 (1952), 46–54.

McKie, D. and G. R. de Beer (b), 'Newton's apple – an addendum', *Notes and Records of the Royal Society* 9 (1952), 333–5.

McKillop, Alan Dugald. *The Background of Thomson's* Seasons (Minneapolis: University of Minnesota Press, 1942).

McKitterick, David. 'Introduction', in David McKitterick (ed.), *The Making of the Wren Library* (Cambridge: Cambridge University Press, 1995), pp. 1–27.

McManners, John. *Death and the Enlightenment: Changing Attitudes to Death among Christians and Unbelievers in Eighteenth-century France* (Oxford: Clarendon Press, 1981).

McNeil, Maureen. 'Newton as national hero', in John Fauvel *et al.* (eds), *Let Newton Be!* (Oxford: Oxford University Press, 1988), pp. 223–39.

McSwiny, Owen. *To the Ladies and Gentlemen of Taste in Great-Britain and Ireland* (London, 1720?).

Meek, Ronald L. *Turgot on Progress, Sociology and Economics* (Cambridge: Cambridge University Press, 1973).

Mercier, Louis-Sébastien. *Le Génie, le goût et l'esprit* (The Hague, 1756).

Mercier, Louis-Sébastien. *Mon bonnet de nuit* (4 vols, Neuchâtel, 1784–6).

Merton, Robert. *Science, Technology and Society in Seventeenth-century England* (New York: Howard Fertig, 1970).

Merton, Robert. *On the Shoulders of Giants: a Shandean Postscript* (Chicago and London: University of Chicago Press, 1993).

Meyer, Rudolf W. *Leibniz and the Seventeenth-century Revolution*, transl. J. P. Stern (Cambridge: Bowes and Bowes, 1952).

Mialet, Hélène. 'Do angels have bodies? Two stories about subjectivity in science', *Social Studies of Science* 29 (1999), 551–81.

Millar, Oliver. *The Tudor, Stuart and Early Georgian Pictures in the Collection of Her Majesty the Queen* (2 vols, London: Phaidon Press, 1963).

Miller, David Philip. '"Puffing Jamie": the commercial and ideological importance of being a "philosopher" in the case of the reputation of James Watt (1736–1819)', *History of Science* 38 (2000), 1–24.

Milner, Thomas. *The Gallery of Nature* (London: Wm S. Orr & Co., 1846).

Money, David K. *The English Horace: Anthony Alsop and the Tradition of British Latin Verse* (Oxford: Oxford University Press, 1998).

Moore, James R. 'Charles Darwin lies in Westminster Abbey', *Biological Journal of the Linnaean Society* 17 (1982), 97–113.

Morrell, John B. and Arnold Thackray. *Gentlemen of Science: Early Years of the British Association for the Advancement of Science* (Oxford: Clarendon, 1981).

Morrill, John (ed.). *The Oxford Illustrated History of Tudor and Stuart Britain* (Oxford and New York: Oxford University Press 1996).

Morton, Alan Q. and Jane A. Wess. *Public and Private Science: The King George III Collection* (Oxford: Oxford University Press, 1993).

Muirhead, James Patrick. *The Life of James Watt, with Selections from his Correspondence* (London: John Murray, 1859).

Mullan, John. 'Gendered knowledge, gendered minds: Women and Newtonianism, 1690–1760', in Marina Benjamin (ed.) *A Question of Identity: Women, Science and Literature* (New Brunswick, NJ: Rutgers University Press, 1993), pp. 41–56.

Munby, A. N. L. *The Cult of the Autograph Letter in England* (University of London: Athlone Press, 1962).

Munby, A. N. L. *The History and Bibliography of Science in England: The First Phase, 1833–1845* (Berkeley and Los Angeles: University of California Press, 1968).

Munsell, F. Darrell. *The Victorian Controversy Surrounding the Wellington War Memorial: The* Archduke *of Hyde Park Corner* (Lewisten: Edwin Mellen Press, 1991).

Murdoch, Ruth T. 'Newton's law of attraction and the French Enlightenment', PhD dissertation, Columbia University, 1950.

Murdoch, Ruth T. 'Newton and the French Muse', *Journal of the History of Ideas* 19 (1958), 323–34.

Nash, Richard. *John Craige's Mathematical Principles of Christian Theology* (Carbondale and Edwardsville: Southern Illinois University Press, 1991).

Nenadic, Stana. 'Print collecting and popular culture in eighteenth-century Scotland', *History* 82 (1997), 203–22.

Newton, Isaac. *Opticks* (London: G. Bell, 1931).

Newton, Isaac. *Observations upon the Prophecies of Daniel, and the Apocalypse of St John* (Cave Junction: Oregon Institute of Science and Medicine, 1991).

Nicholson, Kathleen. 'The ideology of feminine "virtue": the vestal virgin in French eighteenth-century allegorical portraiture', in Joanna Woodall (ed.), *Portraiture: Facing the Subject* (Manchester and New York: Manchester University Press, 1997), 52–72.

Nicolson, Marjorie Hope. *Newton Demands the Muse: Newton's* Opticks *and the Eighteenth-century Poets* (Princeton: Princeton University Press, 1946).

Nochlin, Linda. *Women, Art, Power and Other Essays* (London: Thames and Hudson, 1989).

Nolhac, Pierre de. *La vie et l'œuvre de Maurice Quentin de la Tour* (Paris: H. Piazza, 1930).

Nora, Pierre (ed.). *Les Lieux de mémoire* (3 vols, Paris: Gallimard, 1997).

Noyes, Alfred. *The Torch-bearers* (3 vols, Edinburgh: William Blackwood, 1926–30).

Nurmi, Martin N. 'Blake's "Ancient of Days" and Motte's frontispiece to Newton's *Principia*', in Vivian de Sola Pinto (ed.), *The Divine Vision: Studies in the Poetry and Art of William Blake* (London: Victor Gollanz, 1957).

O'Dwyer, Frederick. *The Architecture of Deane and Woodward* (Cork: Cork University Press, 1997).

Okri, Ben. 'Newtonian enigmas', *Times Saturday Review*, 4 July 1992, 41.

Olson, Richard. *Science Deified and Science Defied: The Historical Significance of Science in Western Culture*, vol. 2 (Berkeley, Los Angeles, Oxford: University of California Press, 1990).

Osler, Margaret J. 'Mixing metaphors: science and religion or natural philosophy and theology in early modern Europe', *History of Science* 36 (1998), 91–113.

Outram, Dorinda. 'The language of natural power: the "Éloges" of Georges Cuvier and the public knowledge of nineteenth-century science', *History of Science* 16 (1978), 153–78.

Ozouf, Mona. 'Le Panthéon: l'école normale des morts', in Pierre Nora (ed.), *Les Lieux de mémoire* (3 vols, Paris: Gallimard, 1997), vol. 1, pp. 155–78.

Pahin-Champlain de la Blancherie, Mammès Claude Catherine. *De par toutes les nations* (London: W. & C. Spilsburg, 1796).

Paradis, James G. 'Satire and science in Victorian culture', in Bernard Lightman (ed.), *Victorian Science in Context* (Chicago and London: University of Chicago Press, 1997), pp. 143–75.

Patterson, Helen Temple. 'Les grands poètes romantiques français et Mercier (Hugo, Vigny, Lamartine, Musset et Gautier)', in Hermann Hofer (ed.), *Louis-Sébastien Mercier: précurseur et sa fortune.* (Munich: Wilhelm Fink Verlag, 1977), pp. 197–246.

Pattison, William. *Poetical Works* (London: for H. Curll, 1728).

Paulson, Ronald. *Hogarth's Graphic Works* (2 vols, New Haven: Yale University Press, 1970).

Paulson, Ronald. *Emblem and Expression: Meaning in English Art of the Eighteenth Century* (London: Thames and Hudson, 1975).

Paulson, Ronald. *Hogarth: High Art and Low, 1732–1750* (Cambridge: Lutterworth Press, 1991).

Pearson, Roger. *The Fables of Reason: A Study of Voltaire's* Contes philosophiques (Oxford: Clarendon Press, 1993).

Pemberton, Henry. *A View of Sir Isaac Newton's Philosophy* (London, 1728).

Pérez-Gómez, Alberto. *Architecture and the Crisis of Modern Science* (Cambridge, MA, and London: MIT Press, 1983).

Pérouse de Montclos, Jean-Marie. *Étienne-Louis Boulleé (1728–1799): de l'architecture classique à l'architecture révolutionnaire* (Paris: Arts et Métiers Graphiques, 1969).

Pérouse de Montclos, Jean-Marie. '"De nova stelli anni 1784"', *Revue de l'Art* 58–9 (1983), 75–84.

Pérouse de Montclos, Jean-Marie. *Étienne-Louis Boullée* (Paris: Flammarion, 1994).

Physick, John. *Designs for English Sculpture 1680–1860* (London: HMSO, 1969).

Picon, Antoine. *French Architects and Engineers in the Age of Enlightenment*, transl. Martin Thom (Cambridge: Cambridge University Press, 1992).

Pietas Academiæ Oxiensis in obitum augustissimæ et desideratissimæ reginæ Carolinæ (Oxford, 1738).

Place, Pierre-Antoine de la. *Recueil d'épitaphes* (3 vols, Brussels, 1782).

Pointon, Marcia. *Hanging the Head: Portraiture and Social Formation in Eighteenth-century England* (New Haven and London: Yale University Press, 1993).

Pointon, Marcia. 'Shakespeare, portraiture and national identity', *Shakespeare Jahrbuch* 133 (1997), 29–53.

Poole, Robert. *Time's Alteration: Calendar Reform in Early Modern England* (London and Bristol, PA: UCL Press, 1998).

Pope, Alexander. *Works* (Ware: Wordsworth, 1995).

Popkin, Richard H. 'Newton and Fundamentalism, II,' in James E. Force and Richard H. Popkin (eds), *Essays on the Context, Nature, and Influence of Isaac Newton's Theology* (Dordrecht, Boston and London: Kluwer Academic, 1990), pp. 165–80.

Popkin, Richard H. 'Newton and the origins of Fundamentalism', in Edna Ullmann-Margalit (ed.), *The Scientific Enterprise* (The Bar-Hillel Colloquium: Studies in History, Philosophy, and Sociology of Science, volume 4) (Dordrecht, Boston and London: Kluwer Academic Publishers, 1992), pp. 241–59.

Priestley, Joseph. *A Description of a Chart of Biography; With a Catalogue of All the Names Inserted in It, and the Dates Annexed to Them* (Warrngton, 1765).

Priestley, Joseph. *The History and Present State of Electricity, with Original Experiments* (London: for J. Dodsley, 1767).

Prowett, Charles Gipps. 'The English universities and their reforms', *Blackwood's Edinburgh Magazine* 65 (1849), 235–44.

Pupil, François. 'La vogue des célébrités sculptées dans le contexte historiographique et littéraire', in Daniel Rabreau and Bruno Tollon (eds), *Le Progrès des arts réunis 1763–1815: mythe culturel, des origines de la Révolution à la fin de l'Empire?* (Bordeaux: CERCAM, 1992), pp. 317–27.

Pye, Charles. *Provincial Copper Coins or Tokens, Issued between the Years 1787 and 1796* (London: John Nichols, 1796).

Quinault, Roland. 'The cult of the centenary, c. 1784–1914', *Historical Research* 71 (1998), 303–23.

Racine, Louis. *La Religion* (Paris: Desaint at Saillant, 1763).

Rackham, Bernard. *Catalogue of the Glaisher Collection of Pottery and Porcelain in the Fitzwillliam Museum Cambridge* (2 vols, Cambridge: Antique Collectors Club, 1987).

Ralph, John. *A Critical Review of the Publick Buildings, Statues and Ornaments in, and about London and Westminster* (London: C. Ackers, 1734).

Raspe, R. E. *A Descriptive Catalogue of a General Collection of Ancient and Modern Engraved Gems, Cameos as well as Intaglios, Taken from the Most Celebrated Cabinets in Europe; and Cast in Coloured Pastes, White Enamel, and Sulphur, by James Tassie, modeller* (2 vols, London: James Tassie, 1791).

Reid, Thomas. *An Inquiry into the Human Mind on the Principles of Common Sense*, ed. Derek R. Brookes (Edinburgh: Edinburgh University Press, 1997).

Reilly, Robin. *Wedgwood Portrait Medallions: An Introduction* (Barrie and Jenkins, London, 1973).

Reily, John and W. K. Wimsatt, 'A Supplement to *The Portraits of Alexander Pope*', in René Wellek and Alvaro Ribeiro (eds), *Evidence in Literary Scholarship: Essays in Memory of James Marshall Osborn* (Oxford: Clarendon Press, 1979), pp. 123–64.

Ricard, Dominique. *La Sphère, poëme en huit chants* (Paris: Le Clerc, 1796).

Rice, Adrian. 'Augustus de Morgan: historian of science', *History of Science* 34 (1996), 201–40.

Richardson, George. *Iconology* (2 vols, New York and London: Garland, 1979 facsimile reprint of 1779 edition).

Roe, Nicholas. *John Keats and the Culture of Dissent* (Oxford: Clarendon Press, 1997)

Romney, John. *Memoirs of the Life and Works of George Romney* (London: Baldwin & Cradock, 1830).

Rose, Mark. *Authors and Owners: The Invention of Copyright* (Cambridge, MA, and London: Harvard University Press, 1993).

Rosenau, Helen. *Social Purpose in Architecture: Paris and London Compared, 1760–1800* (London: Studio Vista, 1970).

Roucher, Jean-Antoine. *Les Mois, poème en douze chants* (2 vols, Paris: Quillau, 1779).

Rousseau, George S. 'Mysticism and millenarianism: "Immortal Dr Cheyne"', in Richard Popkin (ed.), *Millenarianism and Messianism in English Literature and Thought 1650–1800* (Leiden: E. J. Bull, 1988), pp. 81–126.

Rowbottom, Margaret E. 'John Theophilus Desaguliers (1683–1744)', *Proceedings of the Huguenot Society of London* 21 (1968), 196–218.

Russell, Bertrand. *The Analysis of Matter* (New York: Dover, 1954).

Russell, Bertrand. 'Recent work on the philosophy of Leibniz', in Harry G. Frankfurt (ed.), *Leibniz: A Collection of Critical Essays* (New York: Anchor Books, 1972), pp. 365–400.

Saint-Lambert, Jean François de. *Les Saisons, poème* (Amsterdam, 1771).

Saint-Simon, Claude Henri de. *Lettres d'un habitant de Genève à ses contemporains* (Paris: Félix Alcan, 1925).

Samuel, Raphael. *Theatres of Memory: Past and Present in Contemporary Culture* (London and New York Verso, 1994).

Schaffer, Simon. 'Scientific discoveries and the end of natural philosophy', *Social Studies of Science* 16 (1986), 387–420.

Schaffer, Simon. 'Priestley and the politics of spirit', in R. G. W. Anderson and Christopher Lawrence (eds), *Science, Medicine and Dissent: Joseph Priestley (1733–1804)* (London: The Science Museum, 1987), pp. 39–53.

Schaffer, Simon. 'Glass works: Newton's prisms and the uses of experiment', in David Gooding, Trevor Pinch and Simon Schaffer (eds), *The Uses of Experiment: Studies in the Natural Sciences* (Cambridge: Cambridge University Press, 1989).

Schaffer, Simon. 'Genius in Romantic natural philosophy', in Andrew Cunningham and Nicholas Jardine (eds), *Romanticism and the Sciences* (Cambridge: Cambridge University Press, 1990).

Schaffer, Simon. 'Halley, Delisle, and the making of the comet', in

Norman J. W. Thrower (ed.), *Standing on the Shoulders of Giants: A Longer View of Newton and Halley* (Berkeley, Los Angeles and Oxford: University of California Press, 1990), pp. 254–98.

Schaffer, Simon. 'Newtonianism', in R. C. Olby *et al.* (eds), *Companion to the History of Modern Science* (London and New York: Routledge, 1990), pp. 610–26.

Schaffer, Simon. 'The earth's fertility as a social fact in early modern Britain', in M. Teich and R. Porter (eds), *Nature and Society in Historical Context* (Cambridge: Cambridge University Press, 1997), pp. 124–47.

Schweber, S. S. 'John Herschel and Charles Darwin: a study in parallel lives', *Journal of the History of Biology* 22 (1989), 1–71.

Searby, Peter. *A History of the University of Cambridge*, vol. 3, *1750–1870* (Cambridge: Cambridge University Press, 1997).

Secord, Anne. 'Science in the pub: artisan botanists in early nineteenth-century Lancashire', *History of Science* 32 (1994), 269–315.

Secord, James A. 'Newton in the nursery: Tom Telescope and the philosophy of tops and balls, 1761–1838', *History of Science* 23 (1985), 127–51.

Secord, James A. *Victorian Sensation: The Extraordinary Publication, Reception, and Secret Authorship of* Vestiges of the Natural History of Creation (Chicago and London: University of Chicago Press, 2000).

Sepper, Dennis L. 'Goethe against Newton: towards saving the phenomena', in *Goethe and the Sciences: A Reappraisal* (Dordrecht: D. Reidel, 1987), pp. 175–218 (reprinted from *Inquiry* 15 (1972), 363–86).

Shapin, Steven. 'Of gods and kings: natural philosophy and politics in the Leibniz–Clarke disputes', *Isis* 72 (1981), 187–215.

Shapin, Steven. 'Robert Boyle and mathematics: reality, representation, and experimental practice', *Science in Context* 2 (1988), 23–58.

Shapin, Steven. 'Who was Robert Hooke?', in Michael Hunter and Simon Schaffer (eds), *Robert Hooke: New Studies* (Woodbridge: Boydell Press, 1989), pp. 253–85.

Shapin, Steven. 'The philosopher and the chicken: on the dietetics of disembodied knowledge', in Christopher Lawrence and Steven Shapin (eds), *Science Incarnate: Historical Embodiments of Natural Knowledge* (Chicago and London: University of Chicago Press, 1998), pp. 21–50.

Sharpe, William. *A Dissertation upon Genius (1755)*, ed. William Bruce Johnson (New York: Scholar's Facsimiles and Reprints, 1973).

Shaw, George Bernard. *In Good King Charles's Golden Days* (London: Constable, 1939).

Shawe-Taylor, Desmond. *Genial Company: The Theme of Genius in Eighteenth-century British Portraiture* (Nottingham: Nottingham University Art Gallery, 1987).

Shawe-Taylor, Desmond. *The Georgians: Eighteenth-century Portraiture and Society* (London: Barrie and Jenkins, 1990).

Shell, Alison. 'The antiquarian satirized: John Clubbe and the *Antiquities of Wheatfield*,' in Arnold Hunt, Giles Mandelbrote and Alison Shell (eds), *The Book Trade and its Customers 1450–1900* (Winchester: St Paul's Bibliographies, 1997), pp. 223–45.

Shelley, Percy Bysshe. *The Letters of Percy Bysshe Shelley*, ed. Frederick L. Jones (2 vols. Oxford: Clarendon Press, 1964).

Sherlock, Martin. *Letters from an English Traveller, translated from the French original printed at Geneva and Paris* (London: J Nichols *et al.*, 1780; New York: Garland, 1971).

Shortland, Michael and Richard Yeo, 'Introduction', in Michael Shortland and Richard Yeo (eds), *Telling Lives in Science: Essays on Scientific Biography* (Cambridge: Cambridge University Press, 1996), pp. 1–44.

Simon, Robin. *The Portrait in Britain and America with a Biographical Dictionary of Portrait Painters 1680–1914* (Oxford: Phaidon, 1987).

Simoni, Anna E. C. 'Newton in the timberyard: the device of Frans Houttuyn, Amsterdam', *British Library Journal* 1 (1975), 84–9.

Sir Isaac Newton: Man of Science and Officer of the Royal Mint (London: Royal Mint, 1992).

Smart, Christopher. *Selected Poems* (London: Penguin, 1990).

Smernoff, Richard A. *André Chénier* (Boston: Twayne Publishers, 1977).

Smiles, Samuel. *Self-help; with Illustrations of Character and Conduct* (London: John Murray, 1859).

Smith, Charles John. *Historical and Literary Curiosities, Consisting of Facsimiles of Original Documents* (London: Henry G. Bohn, 1840).

Smith, David Eugene. *The Portrait Medals of Sir Isaac Newton* (Boston: Ginn & Co., 1912).

Smith, David Eugene. 'Portraits of Sir Isaac Newton', in W. J. Greenstreet (ed.), *Isaac Newton 1642–1727* (London: G. Bell and Sons, 1927), pp. 171–81.

Smith, Earl L. *Yankee Genius: A Biography of Roger W. Babson – Pioneer in Investment Counseling and Business Forecasting Who Capitalized on Investment Patience* (New York: Harper, 1954).

Smith, John Thomas. *An Antiquarian Ramble in the Streets of London, with Anecdotes of their Most Celebrated Residents* (2 vols, London: Richard Bentley, 1846).

Smith, Logan Pearsall. *Four Words: Romantic, Originality, Creative, Genius* (London: Clarendon Press, 1924).

Snelling, Thomas. *Thirty-three Plates of English Medals* (London: for Thomas Snelling, 1776).

Snobelen, Stephen D. 'On reading Isaac Newton's *Principia* in the 18th century', *Endeavour* 22 (1998), 159–63.

Snobelen, Stephen D. 'Isaac Newton, heretic: the strategies of a Nicodemite', *British Journal for the History of Science* 32 (1999), 381–49.

Sorel, Philippe. 'Trois sculptures de l'époque révolutionnaire: propositions d'attributions', *Gazette des Beaux-Arts* 116 (1990), 137–44.

Sorrenson, Richard. 'Dollond and Son's pursuit of achromaticity, 1758–1789', unpublished paper.

Sotheby, William. *Lines Suggested by the Third Meeting of the British Association for the Advancement of Science, Held at Cambridge, in June, 1833* (London: G. & W. Nicol and J. Murray, 1834).

Spadafora, David. *The Idea of Progress in Eighteenth-century Britain* (New Haven and London: Yale University Press, 1990)

Spence, Joseph. *Observations, Anecdotes, and Characters of Books and Men Collected from Conversation*, ed. James M Osborn (2 vols, Oxford: Clarendon Press, 1966).

Spencer, M. C. *Charles Fourier* (Boston: Twayne Publishers, 1981).

Spencer, T. J. B. 'Lucretius and the scientific poem in English', in *Lucretius*, ed. D. R. Dudley (London: Routledge & Kegan Paul, 1965), pp. 131–64.

Stafford, Barbara Maria. 'Science as fine art: another look at Boullée's *Cenotaph for Newton*', *Studies in Eighteenth-Century Culture* 11 (1982), 241–78.

Stafford, Fiona. *The Last of the Race: The Growth of a Myth from Milton to Darwin* (Oxford: Clarendon Press, 1994).

Staley, Richard. 'On the histories of relativity: the propagation and elaboration of relativity theory in participant histories in Germany, 1905–11', *Isis* 89 (1998), 263–99.

Starobinski, Jean. *1789: les emblèmes de la raison* (Paris: Flammarion, 1973).

Stevens, George Alexander. *Works* (London: James Cundee, 1807).

Stewart, J. Douglas. *Godfrey Kneller* (London: G. Bell & Sons, 1971).

Stewart, J. Douglas. *Sir Godfrey Kneller and the English Baroque Portrait* (Oxford: Clarendon Press, 1983).

Stewart, Larry. *The Rise of Public Science: Rhetoric, Technology, and Natural Philosophy in Newtonian Britain, 1660–1750* (Cambridge: Cambridge University Press, 1992).

Stewart, Larry. 'Other centres of calculations, or, where the Royal Society didn't count: commerce, coffee-houses and natural philosophy in early modern London', *British Journal for the History of Science* 32 (1999), 133–53.

Stukeley, William. *Memoirs of Sir Isaac Newton's Life, Being Some Account of His Family and Chiefly of the Junior Part of His Life*, ed. A. Hastings White (London: Taylor and Francis, 1936).

Suzuki, Keiko. '*Yokohama-e* and *Kaika-e* prints: Japanese interpretations of self and other from 1860 through the 1880s', in Helen Hardacre and Adam L. Kern (eds), *New Directions in the Study of Meiji Japan*, (Leiden, New York and Cologne: Brill, 1997), pp. 676–87.

Symonds, R. W. 'A picture machine of the 18th century', *Country Life* 96 (25 August 1944), 336–7.

Tallis, Raymond. *Newton's Sleep: The Two Cultures and the Two Kingdoms* (Basingstoke: Macmillan, 1995).

Taylor, Charles. *Sources of the Self: The Making of the Modern Identity* (Cambridge: Cambridge University Press, 1989).

Taylor, Samuel. 'Artists and *philosophes* as mirrored by Sèvres and Wedgwood', in Francis Haskell, Anthony Levi and Robert Shackleton (eds), *The Artist and the Writer in France: Essays in Honour of Jean Seznec* (Oxford: Clarendon Press, 1974), pp. 21–39.

Telescope, Tom. *The Newtonian Philosophy, and Natural Philosophy in General, Explained and Illustrated by Familiar Objects, in a Series of Entertaining Lectures* (London: Thomas Tegg, 1838).

Terrall, Mary. 'Émilie du Châtelet and the gendering of science', *History of Science* 33 (1995), 283–310.

Terrall, Mary. 'Gendered spaces, gendered audiences: inside and outside the Paris Academy of Sciences', *Configurations* 2 (1995), 207–32.

Thackray, Arnold. '"The business of experimental philosophy": the early Newtonian group at the Royal Society', *Actes du XIIe Congrès International d'Histoire des Sciences* 3B (1970–71), 155–9.

The European Fame of Isaac Newton, catalogue of 1974 exhibition at the Fitzwilliam Museum, Cambridge.

The Man at Hyde Park Corner: Sculpture by John Cheere 1709–1787 (Leeds, 1974).

Theerman, Paul. 'Unaccustomed role: the scientist as historical biographer – two nineteenth-century portrayals of Newton', *Biography* 8 (1985), 145–62.

Thomas, Keith. *Man and the Natural World: Changing Attitudes in England 1500–1800* (Harmondsworth: Penguin, 1983).

Thomas, W. K. and Warren U. Ober. *A Mind for Ever Voyaging: Wordsworth at Work Portraying Newton and Science* (Edmonton: University of Alberta Press, 1989).

Thomson, James. *Works* (2 vols, London: for A. Millar, 1738).

Thrale, Hester Lynch. *Thraliana: The Diary of Mrs Hester Lynch Thrale, 1776–1809*, ed. Katherine Balderston (2 vols, Oxford: Oxford University Press, 1942).

Timbs, John. *School-days of Eminent Men* (London: Lockwood & Co., 1862).

Tollet, Elizabeth. *Poems on Several Occasions* (London: for T. Lownds, c.1760).

Toole, Betty A. *Ada, the Enchantress of Numbers: A Selection from the Letters of Lord Byron's Daughter and Her Description of the First Computer* (Mill Valley, CA: Strawberry Press, 1992).

Traweek, Sharon. *Beamtimes and Lifetimes: The World of High Energy Physicists* (Cambridge, MA, and London: Harvard University Press, 1988).

Tupper, Martin Farquahar. *Three Hundred Sonnets* (London: Arthur Hall, Virtue & Co., 1860).

Turnbull, H. W. *et al. The Correspondence of Isaac Newton* (7 vols, Cambridge: Cambridge University Press, 1959–77).

Turner, Frank M. 'Victorian scientific naturalism', *Victorian Studies* 18 (1975), 325–43, reprinted in *Contesting Cultural Authority: Essays in Victorian Intellectual Life* (Cambridge: Cambridge University Press, 1993), pp. 131–50.

Turnor, Edmund. *Collections for the History of the Town and Soke of Grantham* (London: William Miller, 1806).

Uglow, Jenny. *Hogarth: A Life and a World* (London: Faber and Faber, 1997).

Vago, A. L. *Orthodox Phrenology* (London: Simpkin, Marshall & Co., 1871).

Vartanian, Aram. 'Diderot et Newton', in Claude Blanckaert, Jean-Louis Fischer and Roselyne Rey (eds), *Nature, histoire, société: essais en hommage à Jacques Roger* (Paris: Klincksieck, 1995), pp. 61–77.

Vicinus, Martha. '"Tactful organising and executive power": biographies of Florence Nightingale for girls', in Michael Shortland and Richard Yeo (eds), *Telling Lives in Science: Essays on Scientific Biography* (Cambridge: Cambridge University Press, 1996), pp. 195–213.

Vidal, Mary. 'David among the moderns: art, science, and the Lavoisiers', *Journal of the History of Ideas* 56 (1995), 595–623.

Vidler, Anthony. *The Writing of the Walls: Architectural Theory in the Late Enlightenment* (London: Butterworth Architecture, 1987).

Vidler, Anthony. *Claude-Nicolas Ledoux: Architecture and Social Reform at the End of the Ancien Régime* (Cambridge, MA, and London: MIT Press, 1990).

Vidler, Anthony. *The Architectural Uncanny: Essays in the Modern Unhomely* (Cambridge, MA, and London: MIT Press, 1992).

Villamil, R. de. *Newton: The Man* (London: Gordon D. Knox, 1931).

Voltaire. *Letters on England*, transl. L. Tancock (Harmondsworth: Penguin 1980).

Walsh, Linda. 'The expressive face: manifestations of sensibility in eighteenth-century French art', *Art History* 19 (1996), 523–50.

Walters, Robert L. 'The allegorical engravings in the Ledet-Lesbordes edition of the *Élemens de la philosophie de Newton*', in R. J. Howells *et al.* (eds), *Voltaire and His World: Studies Presented to W. H. Barber* (Oxford: Voltaire Foundation, 1985) pp. 27–49.

Ward, Humphry and W. Roberts. *Romney: A Biographical and Critical Essay with a Catalogue Raisonné of his Works* (2 vols, London: Thomas Agnew & Sons, 1904).

Watkins, Eric. 'The laws of motion from Newton to Kant', *Perspectives on Science* 5 (1997), 311–48.

Watt, Ian. *Myths of Modern Individualism: Faust, Don Quixote, Don Juan, Robinson Crusoe* (Cambridge: Cambridge University Press, 1996).

Wattenberg, Diedrich. 'Eine Totenmaske Isaac Newtons?', *Internationale*

Zeitschrift für Geschichte der Naturwissenschaften, Technik und Medizin 14 (1977) 1, 43–6.

Webb, M. I. 'Busts of Sir Isaac Newton', *Country Life* 111 (25 January 1952), 216–18.

Webb, M. I. *Michael Rysbrack: Sculptor* (London: Country Life, 1954).

Webber, Roger Babson. *A Descriptive Catalog of the Grace K. Babson Collection of the Works of Sir Isaac Newton* (New York: Herbert Reichner, 1950).

Webster, Mary. 'Taste of an Augustan collector: The collection of Dr Richard Mead – I', *Country Life* 147 (29 Januiary 1970), 249–51.

Wells, George A. *Goethe and the Development of Science, 1750–1900* (Alphen aan den Rijn: Sijthoff & Noordhoff, 1978).

Werrett, Simon. 'Newton in eighteenth-century Russia', unpublished paper.

Werskey, Gary. *The Visible College: A Collective Biography of British Scientists and Socialists of the 1930s* (London: Free Association Books, 1988).

Westfall, Richard S. 'Newton and the fudge factor', *Science* 179 (1973), 751–8.

Westfall, Richard S. *Never at Rest: A Biography of Isaac Newton* (Cambridge: Cambridge University Press, 1980).

Westfall, Richard S. 'Newton and his biographer', in Samuel H. Baron and Carl Pletch (eds), *Introspection in Biography: The Biographer's Quest for Self-awareness* (Hillsdale, NJ: The Analytic Press, 1985), pp. 175–89.

Whewell, William. *A History of the Inductive Sciences* (3 vols, London: John W. Parker, 1857).

White, Michael. *The Last Sorcerer* (London: Fourth Estate, 1997).

White, Walter. *The Journals of Walter White* (London: Chapman and Hall, 1898).

Whitwell, F. 'Popular science', *Quarterly Review* 84 (1848–9), 307–44.

Wiebenson, Dora. *The Picturesque Garden in France* (Princeton: Princeton University Press, 1978).

Wilbert, Chris. 'The apple falls from grace', in Murray Bookchin *et al.* (eds), *Deep Ecology and Anarchism: A Polemic* (London: Freedom Press, 1993).

Wilde, Christopher B. 'Hutchinsonianism, natural philosophy and religious controversy in eighteenth-century Britain', *History of Science* 18 (1980), 1–24.

Williams, Carolyn D. *Pope, Homer, and Manliness: Some Aspects of Eighteenth-century Classical Learning* (London and New York: Routledge, 1993).

Williams, Perry. 'Passing on the torch: Whewell's philosophy and the principles of English university education', in Menaschem Fisch and Simon Schaffer (eds), *William Whewell: A Composite Portrait* (Oxford: Clarendon Press, 1991), pp. 117–47.

Williams, Raymond. *Keywords: A Vocabulary of Culture and Society* (New York: Oxford University Press, 1976).

Williamson, Karina. 'Smart's *Principia*: science and anti-science in *Jubilate Agno*', *Review of English Studies* 30 (1979), 409–22.

Willis, Peter. *Charles Bridgeman and the English Landscape Garden* (London: A. Zwemmer, 1977).

Wilson, Michael I. *William Kent: Architect, Designer, Painter, Gardener, 1685–1748* (London: Routledge and Kegan Paul, 1984).

Wimsatt, William Kurtz. *The Portraits of Alexander Pope* (New Haven and London: Yale University Press, 1965).

Winstanley, D. A. *Early Victorian Cambridge* (Cambridge: Cambridge University Press, 1940).

Winter, Jay. *Sites of Memory, Sites of Mourning: the Great War in European Cultural History* (Cambridge: Cambridge University Press, 1995).

Wolf, Edwin. *The Library of James Logan of Philadelphia 1674–1751* (Philadelphia: Library Company of Philadelphia, 1974).

Wollstonecraft, Mary. *Vindication of the Rights of Women* (Harmondsworth: Penguin, 1975).

Woodmansee, Martha. 'The genius and the copyright: economic and legal conditions of the emergence of the "author"', *Eighteenth-century Studies* 17 (1984), 425–48.

Woodward, Kenneth. *Making Saints: Inside the Vatican: Who Become Saints, Who Do Not, and Why . . .* (London: Chatto and Windus, 1991).

Wright, Patrick. *On Living in an Old Country: The National Past in Contemporary Britain* (London: Verso, 1985).

Wright, T. R. *The Religion of Humanity: The Impact of Comtean Positivism on Victorian Britain* (Cambridge: Cambridge University Press, 1986).

Yarrington, Alison. *The Commemoration of the Hero 1800–1864: Monuments to the British Victors of the Napoleonic Wars* (New York and London: Garland, 1988).

Yeo, Richard. 'Scientific method and the image of science 1831–1891', in

Roy MacLeod and Peter Collins (eds), *The Parliament of Science: The British Association for the Advancement of Science 1831–1981* (Northwood, Middx: Science Reviews Ltd, 1981), pp. 65–88.

Yeo, Richard. 'An idol of the market-place: Baconianism in nineteenth-century Britain', *History of Science* 23 (1985), 251–98.

Yeo, Richard. 'Genius, method and morality: images of Newton in Britain, 1760–1860', *Science in Context* 2 (1988), 257–84.

Yeo, Richard. *Defining Science: William Whewell, Natural Knowledge, and Public Debate in Early Victorian Britain* (Cambridge: Cambridge University Press, 1993).

Yeo, Richard. 'Alphabetical lives: scientific biography in historical dictionaries and encyclopaedias', in Michael Shortland and Richard Yeo (eds), *Telling Lives in Science: Essays on Scientific Biography* (Cambridge: Cambridge University Press, 1996), pp. 139–69.

Yeo, Richard. *Encyclopaedic Visions: Scientific Dictionaries and Enlightenment Culture* (Cambridge: Cambridge University Press, 2001).

Yolton, John W. *Thinking Matter: Materialism in Eighteenth-century Britain* (Oxford: Basil Blackwell, 1983).

Young, Edward. *Conjectures on Original Composition* (London: for A. Millar, 1759).

Zinsser, Judith P. 'Émilie du Châtelet: Genius, gender, and intellectual authority', in Hilda L. Smith (ed.), *Women Writers and the Early Modern British Political Tradition* (Cambridge: Cambridge University Press, 1998), pp. 168–90.

Zinsser, Judith P. 'Translating Newton's *Principia*: the Marquise du Châtelet's revisions and additions for a French audience', *Notes and Records of the Royal Society* 55 (2001), pp. 227–45.

Ziolkowski, Theodore. *German Romanticism and its Institutions* (Princeton: Princeton University Press, 1990).

Index